PHEROMONES OF SOCIAL BEES

PHEROMONES
OF SOCIAL BEES

John B. Free

Bee Research Unit
University of Wales at Cardiff

Comstock Publishing Associates
a division of Cornell University Press
ITHACA, NEW YORK

First published 1987 by Cornell University Press

Library of Congress Cataloging-in-Publication Data

Free, John Brand
  Pheromones of social bees.

  Bibliography: p.
  Includes indexes.
  1. Bees——Behavior. 2. Bees——Physiology. 3. Insect
hormones. 4. Insect societies. 5. Pheromones. I. Title.
QL569.4.F74   1987   595.79′90451   86-24227
ISBN 0-8014-2004-0

Printed in Great Britain

For Nancy
For many reasons

CONTENTS

PREFACE

The known pheromones of most non-social insects are concerned primarily with mating. In contrast, social insects which need to collaborate in food gathering, brood rearing, colony growth, reproduction and defence employ many different pheromone systems which control almost all of their activities.

Because of the economic importance and fascination of social bees their pheromones have been among the most studied. This is especially true for the honeybee *Apis mellifera* which is native to Africa, Europe and West Asia and has been introduced by man to North and South America, Australasia and many countries in Central and East Asia. Unless stated otherwise the work I have described refers to this species. Unfortunately, so far only fragmentary studies have been made on the three other honeybee species, all of which occur in South East Asia.

Some interesting pheromone discoveries have been made on the stingless bees, particularly on their trail and alarm pheromones, and on nest recognition by primitively social bees. These will be included in the appropriate chapters. Pheromone studies on the social organization of bumblebees and especially on their mating behaviour are so advanced as to warrant chapters on their own.

Attempted economic use of synthetic pheromones of non-social insects is usually limited to monitoring a pest population or directly suppressing a population. In association with the much greater complexity of function and diversification of use of honeybee pheromones, the possibilities of using synthetic pheromones to control bees' activities and increase their efficiency as crop pollinators and honey producers are enormously varied and challenging.

Many of the methods and techniques used in commercial beekeeping were, in fact, developed by the beginning of this century. By their use the behaviour of bees is adapted to some extent to meet beekeeping requirements. But there have been few advances in altering bee behaviour and so directing a colony's activities towards greater productivity. Hopefully, these advances will eventually be achieved by presenting appropriate synthetic chemicals, analogous to the pheromone itself, that will release the required behaviour patterns. Towards the end of appropriate chapters I have attempted to suggest ways in which synthetic pheromones might be used to increase beekeeping efficiency.

Because a single pheromone often has many functions and because more than one pheromone is often involved in a single category of behaviour, it has not been possible to segregate function and pheromone source in the chapter divisions. I have chosen to have chapter divisions based on function as this classification appears to provide a more coherent and complete account. With the exception of the chapter concerned with the orientation functions of the Nasonov pheromone, a number of pheromone sources will be involved in each.

I was most fortunate that my Director of Studies at Jesus College Cambridge was the eminent ethologist Professor W.H. Thorpe FRS. He introduced me to the study of bumblebees and to the writings of F.W.L. Sladen who was the first to appreciate the purpose of the 'scent marks' made by male bumblebees along their flight routes and the functions of the pheromones produced by the Nasonov gland of the honeybee, and was the first to demonstrate the sex attractant of virgin queen honeybees.

I joined the Bee Department, Rothamsted Experimental Station in 1951 to work initially on bumblebees at a time when exciting and important discoveries on honeybee pheromones were beginning to be made. Soon afterwards Nancy W. Speirs, who became my wife, and Dr C.R. Ribbands proved the effectiveness of the honeybee Nasonov pheromone in aiding honeybee orientation to the hive and demonstrated the rapidity with which food collected by foraging honeybees is widely disseminated throughout a colony. They suggested this would provide a means of rapid communication among its members. A year or so later Dr C.G. Butler FRS discovered that the honeybee queen produces a pheromone that is rapidly distributed among the workers of her colony and inhibits them from rearing additional queens. This was the beginning of the queen pheromone studies at Rothamsted in which Dr Butler was joined by Dr R.K. Callow FRS and Dr J. Simpson. It has been my privilege to be engaged in bee research as the bee pheromone story has steadily unfolded and to have among my friends many bee research workers from different parts of the world who have made notable contributions to it.

Especially, it is a pleasure to acknowledge my indebtedness to my former colleagues at Rothamsted Experimental Station with whom I have collaborated at various times to study bee pheromones. These include: Dr C.G. Butler, A.W. Ferguson, Barbara Goult, A. Martin, Dr A. Mudd, Dr J.A. Pickett, R.A. Welch, Jacqueline R. Simpkins, Dr J. Simpson, Dr Yvette Spencer-Booth, P.W. Tomkins and Dr Ingrid H. Williams.

Finally I wish to thank Jacqueline R. Simpkins who has cheerfully helped in many ways throughout the preparation of this book, especially in maintaining and classifying the appropriate reprint collection, and Mrs Jackie Fountain who has willingly and efficiently deciphered my writing and typed the manuscript. To both I am most grateful.

John B. Free

Chapter one

INTRODUCTION

Pheromone uses

A pheromone is a chemical, secreted from the exocrine gland of an animal, that elicits a behavioural or physiological response by another animal of the same species and so acts as a chemical message. It is secreted as a liquid and transmitted as a liquid or gas.

Honeybee colony (*Apis mellifera*) in hollow tree

A pheromone can act directly on the recipient's central nervous system and result in an immediate behavioural response; in such circumstances it is said to have a 'releaser' effect. Or, a pheromone can initiate a physiological change or changes in the recipient which as a result may acquire a new repetoire of behaviour patterns; the pheromone is then said to exhibit a 'primer' effect.

Both effects are apparent among pheromones of *eusocial* bees. These are bees that live together in colonies consisting of individuals of different castes (e.g. queen, worker), with at least some workers being more or less sterile, belonging to at least two generations and co-operating in the care of the young.

Examples of releaser pheromones of eusocial bees are those concerned with sex attraction, alarm and aggression, trail production, clustering and mutual recognition. The primer pheromones of eusocial bees are concerned with

Beekeepers in India examining comb of *Apis cerana* colony in modern hive

An *Apis florea* colony in India

inhibition of reproduction; they have an essential role in the organization and cohesion of the more developed societies.

For eusocial insects, as for non-social insects, the pheromone messages transmitted by males or queens during mating are for members of the opposite sex of their own species, but not necessarily for members of their own colony. In fact many of the behaviour patterns associated with mating tend to favour union between progenies of alien colonies.

The targets of other pheromone messages are usually members of the emitting bee's own colony. There are few exceptions: honeybee alarm pheromones released at the entrance of a hive may deter would-be robber bees from an alien colony and so prevent interspecific fighting and loss of lives; possible appeasement pheromones released by inadvertent intruder bees may deter attacks by guard bees. In most circumstances if pheromone messages reach honeybees of another colony they are either wasted, or, as when the pheromone marks a food or water supply, can even help competing colonies.

There is little evidence of a pheromone language barrier between honeybee colonies although the mandibular gland secretion of queens of different races may have somewhat different proportions of the same components. The pheromone language conveying the primary message is probably common to all honeybee colonies but it may be overlaid with a colony odour that gives it a distinctive flavour. This may either make it meaningless or less acceptable to bees of another colony and may even release a hostile reaction.

An *Apis dorsata* colony in Bangladesh

Methods of transmission

There are a number of ways by which honeybee pheromones are distributed within the colony depending partly on the volatility of the pheromone concerned.

Highly volatile components of low molecular weight, notable among which are the alarm pheromones and attractant pheromones, are transmitted in the air. Bumblebees, stingless bees, *Apis mellifera* and *Apis cerana*, the two honeybee species whose colonies build several parallel combs and are kept in hives, live naturally in enclosed cavities or nests. The ability of bees to renew rapidly the air inside the nest or hive by fanning at the entrance and on the comb could help to remove the volatile pheromones when the need for them is past. Indeed, the sheltered conditions inside a nest would seem to favour communication by airborne pheromones; airborne brood pheromones may help returning foragers to become aware of their colony's food requirements, and airborne queen pheromones could help all bees to be aware of her presence.

In contrast to the other honeybees, colonies of *Apis dorsata*, the great honeybee, and *Apis florea*, the dwarf honeybee, build only single combs and often nest in exposed situations. However, the curtain of worker bees, often several thick, that screens the comb leaves a sheltered cavity beneath it in which the house-bees perform their activities and foragers enter to deposit their loads of nectar and pollen.

Other pheromones functioning within the nest are relatively involatile and are adsorbed onto the body surface where they are preceived by contact chemoperception. Long chain paraffins in the cuticular wax of honeybees tend to act as fixatives. Examples of such pheromones are those responsible for brood recognition, caste regulation and colony odour. They are passed from bee to bee either by direct bodily contact or in the food supply. They may also be deposited and perceived on the surface of the comb.

Pheromones used in the field are usually transmitted in the air. They are used by a swarm or migrating colony when it is clustering and probably also when it is in flight. Pheromones are also used to mark an enemy or a source of forage and attract other bees to it. Some of the pheromones used to mark flowers may be quite persistent.

Bioassays

The development of sound, meaningful bioassays is of paramount importance in the study of bee pheromones, but because of variations in the age, physiological condition and genetic composition of the bees used and the presence of many unknown pheromones it is one of the most difficult tasks confronting the research worker. Indeed, the threshold response of bees to their pheromones and the behaviour released by the different components may well differ with the age and physiological condition of the recipient, the particular task it is undertaking and the behavioural repetoire at its disposal.

Because of the complexities associated with the social behaviour of bees and other environmental parameters, pheromone-induced behaviour may differ markedly between laboratory and field tests. The aim should always be to relate the bioassay as closely as possible to the natural situation. It should also take into account the physiological condition and past experience of the test bees and the concentration of the test chemical. Above all, the test chemical must be presented in the correct behavioural context. It is usually much easier to devise simple quick and efficient bioassays for 'release' than for 'primer' pheromones; bioassays for the latter are often tedious and time consuming.

Many of the apparently conflicting results obtained by different research workers have probably arisen from some of the differences in their bioassay techniques. Indeed, few research workers attempt to copy exactly the bioassay of another; most prefer to develop their own.

The identification and characterization of honeybee pheromones is difficult, and so far none of the pheromonal systems operating within the honeybee colony have been completely simulated by synthetic materials. Nevertheless, much progress has been made, and it is now clear how synthetic pheromones might be used to increase the efficiency of beekeeping, both for honey production and pollination.

Chapter two

COMMUNICATION OF A QUEEN'S PRESENCE

Introduction

The presence of a queen maintains colony cohesion and stability. Her importance is reflected in the attention paid to her by the workers, and in the effect of removing her which first leads to the colony rearing new queens and later, after her prolonged absence, to the development of the workers' ovaries.

Butler (1954a,b) linked two small colonies, one without a queen, by a common arena and found that worker bees from the queenless colony moved into the one with the queen. Other experiments (Free and Spencer-Booth, 1965) have demonstrated that when workers are introduced to colonies with a queen a greater percentage remain than of workers introduced to queenless colonies, especially when no brood is present. Even under highly artificial conditions the queen can greatly influence the behaviour of worker bees. Free and Williams (1973) kept groups of ten bees in small cages, some with and some without tethered queens, for 48 hours before opening the cage entrances and allowing the worker bees to fly; the workers that had been kept with a queen were slower to leave their cages, more of them made orientation flights, more of them returned to their cages and those that did so returned sooner than bees from the queenless cages.

The queen honeybee produces a number of pheromones which jointly attract workers to her, stimulate foraging, brood rearing, comb building and other activities, inhibit queen rearing and the development of workers' ovaries, and so prevent the presence of other reproductive individuals. A knowledge of these pheromones is most important in understanding social cohesion and control within the community, and is an essential preliminary step in controlling reproduction and stimulating foraging.

The pheromones that signal a queen's presence in her colony and which attract workers to her may also have inhibitory or stimulatory effects (Chapters 3, 4 and 6) but their composite functions are far from clear.

The queen's court

The queen's importance and attractiveness is manifest by the special attention paid to her within the nest. When the queen is stationary on the comb she is surrounded by a circle of attendants (known as the queen's 'court' or 'retinue') that face toward her, offer food to her, palpate her with their antennae, and lick her. Workers in the court stand so they are able to move their heads forward and reach the queen with their antennae or tongues; perhaps they are repelled when closer – their behaviour of moving their heads to and fro certainly gives this impression.

Usually eight or more worker attendants are present when a queen is stationary, but they tend to lose contact as she walks rapidly over the comb. For example, an average of 6.2 attendants has been recorded (Allen, 1957) when the queen is moving, increasing to 8.7 when she is laying an egg and 10.8 when she is stationary. The amount young queens are licked and palpated increases as they become mated and lay eggs, but when they have grown old and are laying haploid eggs they receive less attention (Free *et al.*, 1987a). This can be seen in the table below:

Type of queen	Number of workers licking queen	Number of workers palpating queen
Virgin	0.1	4.0
Newly mated	0.3	5.4
One-year-old mated	0.6	7.9
Old mated	0.4	5.7

The presence of a queen seems to be conveyed but a short distance and only those workers within a few millimetres of her react as she passes by. Responses of individual workers to their queen vary greatly. At the queen's approach some workers turn toward her, others appear indifferent and unaware of her, and a few appear to avoid her. More than half the workers that join the court stay less than 30 seconds and three-quarters less than a minute. However, some stay much longer (up to 41 minutes) and try not to be separated from the queen as she moves (Allen, 1957). The reason for these different responses to a queen is unknown, especially as workers that avoid their queen may respond favourably to her presence at other times. A queen's court contains workers from a few hours to 36 days old, but there is no obvious difference in the age of attendants staying short and long periods (Allen, 1955). Perhaps a worker's response to her queen may be related to the length of time elapsed since last in her court so that a worker who has contacted her queen recently is attracted to her little, if at all.

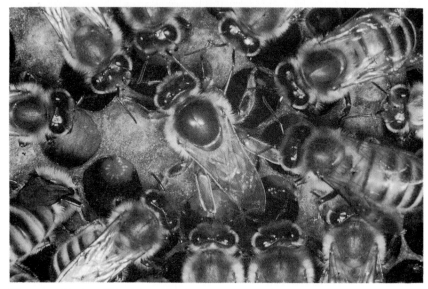

Honeybee (*Apis mellifera*) queen laying an egg while surrounded by her 'court'

The number of bees in the queen's court observed by Allen (1957) was lower during early January (an average of 5.6 workers) than the rest of the year, but was not noticeably low at other times in the winter. However, with diminution in egg laying in late summer, Allen (1957) found that the number of bees licking the queen also diminished, and it remained low throughout the winter.

Free *et al*. (1987a) observed that the number of bees in the queen's court, and the number that licked and palpated the queen, decreased during the course of the winter irrespective of any increase in egg laying, suggesting a decrease in her pheromone production. Perhaps while relatively inactive in a winter cluster, the workers' pheromones requirements are also diminished.

Pain *et al*. (1972) attempted to determine an annual cycle in the emission of queen pheromone by confining the queen at intervals in a filter paper cylinder and then determining the cylinder's attractiveness to worker bees. Unfortunately, the results are difficult to interpret as the amount deposited could depend both on the amount that had accumulated on the queen's integument and on the amount she secreted, but they indicate that maximum production for the year occurred in June.

Pheromone transfer from the queen's court

There is frequent and extensive transfer of food within the honeybee colony (Nixon and Ribbands, 1952; Free, 1957). Food is passed directly from one worker bee to another as well as from workers to drones and the queen. A bee which is begging for food attempts to thrust its tongue between the mouthparts of another bee, and a bee which is offering food opens its mandibles and moves its still folded tongue slightly downwards and forwards from its position of rest; it is conceivable that pheromones are involved as signals in both circumstances. The food given is water, nectar or honey regurgitated from the honey stomach, but it may also on some occasions consist of, or contain, glandular secretions. It was supposed that workers obtained pheromone by licking the queen's body, and that the pheromone became distributed among workers in regurgitated food. Butler (1954a) found that workers seen to lick the queen tended to offer food to other bees within the next five minutes, whereas those that merely palpated her with their antennae did not. He transferred groups of 20 bees from a queenright to a queenless colony at five-minute intervals and found that the queenless part was prevented from rearing queen cells (see page 32); he thought this was because the transferred bees carried queen pheromone in their honey stomachs.

Food transfer. The worker on the right giving food to the worker on the left (*Apis mellifera*)

However, it is doubtful whether pheromones could act chemically after ingestion. There is evidence that if the queen's pheromone is mixed with food containing sugar in concentrations higher than 5–10% it loses its inhibiting effect (Verheijen-Voogd, 1959; Van Erp, 1960; Pain, 1961a,b). Furthermore, even if distribution of pheromones in food does occur, it has become doubtful whether they could be disseminated effectively and quickly enough by this method alone. (Verheijen-Voogd, 1959; Allen, 1965a; Velthuis, 1972; Butler, 1973). For example, Allen (1957) found that at any one time in the summer only about one bee in the court is licking the queen to every twelve examining her with their antennae. Relatively fewer workers lick their queen when she is laying eggs and none when she is moving over the comb (Free *et al.*, 1987a). For example:

Behaviour of queen	Number of workers licking queen	Number of workers palpating queen
Stationary	0.9	9.8
Laying	0.5	7.3
Moving	0	5.2

Instead it seems that queen pheromones are transferred from the queen to workers in her court, and presumably from them to other bees, by physical contact and recognized by chemoperception. Velthuis (1972) showed that a worker that had been confined with a queen was attractive to a small queenless group of workers which treated her as a 'substitute queen', probably because of the queen pheromone adhering to her body.

During transfer of food, the antennae of donor and recipient are continually in motion and often touch each other. Whereas this contact is essential for food transfer to occur (Free, 1956) workers also frequently make antennal contact without accompanying food transfer, and strong circumstantial evidence has accumulated (Free, 1978; Seeley, 1979; Ferguson and Free, 1980; Juska *et al.*, 1981) which indicates that workers in a queen's court obtain queen pheromone on their antennae, and that queen pheromone is distributed through the colony during antennal contact between workers.

Workers palpating and licking their queen's abdomen (*Apis mellifera*)

Workers making antennal contact (*Apis mellifera*)

Workers that have been in a queen's court and palpated her with their antennae have an increased tendency to make reciprocated antennal contact with other workers but it is these other workers that seem mostly to initiate the contact, suggesting that they can detect the presence of an attractant (presumably the queen pheromone) on the antennae of workers from the court and also perhaps on other parts of their bodies to which it has been transferred by personal grooming. Indeed, experiments showed that workers were stimulated to make antennal contact with the excised heads alone of bees from a queen's court, indicating that the presence of queen's pheromone on the antennae, irrespective of the bee's behaviour, is of prime importance. The longer a bee has been in the court, the greater the frequency with which it attracts and is inspected by other workers. The first workers contacted by those leaving the court also have an increased tendency to make reciprocated antennal contacts.

On leaving a court bees range widely throughout the brood area of a colony, frequently contacting other workers. Probably direct contact with the queen stimulates them to greater activity (page 61). However, it is possible that the more active workers are likely to respond to the presence of a nearby queen and enter her court, so that the presence of workers in the court is a consequence of their greater activity and not its cause; information is needed on the activity of bees prior to entering a court.

Within about ten minutes of leaving the queen's court the workers' behaviour and attractiveness returns to that of workers outside the queen's

The worker bee is making antennal contact and begging food from the head alone of another worker (*Apis mellifera*) taken from a queen's court

immediate influence (Fig. 2.1); this is probably because the effects of the stimulation they had received from the queen and the amount of queen pheromone on their antennae has diminished.

Presumably most loss of queen pheromone from the worker's antennae is by volatilization, but we cannot be sure until we have identified the components involved. Continuous air turmoil caused by the movement of thousands of bees in the brood area of a honeybee colony must aid volatilization and dispersal of a pheromone and lead to the need for its

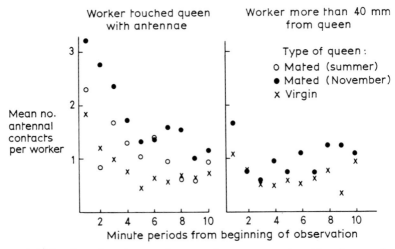

Figure 2.1 Number of reciprocated antennal contacts made with other workers during a ten-minute period by workers immediately after leaving the queen's court and by control workers not near the queen (after Ferguson and Free, 1980).

continued replacement. Ventilation fanning will accelerate the process considerably.

The frequency of antennal contact between workers together with the continued movement of bees would seem to ensure that queen pheromone is adequately dispersed within the colony, particularly among the younger bees which are on the combs containing brood.

There is a greater frequency of antennal contact in colonies with mated queens than in colonies with virgin queens, presumably reflecting the stimulatory effect of an increased queen pheromone level on worker activity. Virgin queens spend a much greater proportion of their time moving over the comb than do mated queens (Free et al., 1987a); because moving queens attract only small courts the amount of pheromone workers obtain from a mobile virgin must be reduced, even though the virgin queen's own movement must aid its distribution to some extent. The fewer antennal contacts made in colonies during the winter, or in colonies headed by failing queens, probably indicates the distribution of less queen pheromone (Free et al., 1987a).

Bees from the queen's court, and the workers first contacted by them on leaving the court, show an increased tendency to participate in food transfer, but the number of contacts involving food transfer are relatively few compared to those involving antennal contact only, and bees from the court are as likely to receive food as to give it. When food transfer does occur it has the desirable effect of prolonging antennal contact.

Bees that lick their queen subsequently make more reciprocated antennal contacts than those that only palpate her with their antennae. This may be because such bees spend a longer time in the court, acquire a greater amount of queen pheromone, and are more attractive to others. However, workers that lick a virgin queen show a pronounced tendency to engage in food transfer with other workers, so the possibility remains that some pheromone communication may occur during food transfer, but if so it usually occurs only in special circumstances and is in general of minor importance and incidental to pheromone transfer by antennal contact.

Therefore it no longer seems likely that queen pheromone has a direct biochemical effect on worker bees after being ingested by them, but instead it seems probable that the queen's presence is communicated by contact chemoperception, followed by appropriate worker behavioural responses or hormonal adjustments. It is possible that bees in the queen's court are themselves stimulated to produce pheromones which induce a reciprocal release of pheromone by workers that contact them and this alone could be a means by which the queens presence is communicated without physical transfer of queen pheromone being necessary. However, the absence of contact other than between the antennae of the workers concerned makes this unlikely.

Pheromone transfer from immature queens

Antennal contact and food transfer are also important in the transfer of pheromones from immature queen honeybees (Free and Ferguson, 1982). Queens are reared in acorn-shaped wax cells that project and hang downwards from the surface of the comb. Workers surrounding queen cells lick and palpate the cell walls. More workers face and palpate open cells containing queen larvae than closed cells containing queen pupae, and the queen larvae themselves are inspected almost continuously. However, workers are particularly inclined to lick queen cells containing pupae.

Immediately after visiting cells containing immature queens, workers engage in prolonged cleaning, particularly of their tongues when they have visited larvae, and of their antennae when they have visited pupae. This intensive grooming may help distribute pheromone over their bodies and increase their attractiveness to other bees. Thereafter, other workers usually initiate and make antennal contact with them.

Presumably the pheromone produced by immature queens is released by diffusion through the walls of the queen cell. Workers can obtain pheromone by direct contact with queen larvae, but they can only obtain queen pupae pheromone that has permeated the queen cell walls. In a queen cell containing a pupa, a space is left between the cocoon and the wax that covers the cell tip (Jay, 1963). At any time during the pupal stage, but usually two or three days before the queen is due to emerge, worker bees remove this wax

Queen cell containing queen larva (*Apis mellifera*)

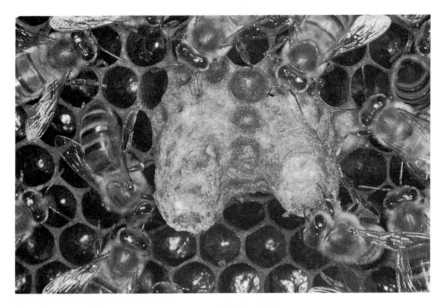

Closed queen cells containing queen pupae (*Apis mellifera*)

tip (Morse and McDonald, 1965). As Boch (1979) suggested, this behaviour may facilitate release of pheromone from the queen cell.

Workers that have visited queen larvae often donate food during antennal contact with other bees, so any pheromone they lick from queen larvae could also be distributed through the colony in food (Free and Ferguson, 1982). There appears to be a division of labour between workers feeding worker larvae and queen larvae, the latter tending to be attended and fed by older bees (Furgala and Boch, 1961; Smith, 1974). Consequently, bees feeding or inspecting queen larvae may be consistent recipients and carriers of relatively large amounts of pheromone. Although workers frequently lick closed queen cells (presumably to obtain pheromone), there is no evidence to suggest that it is subsequently transferred in regurgitated food (Free and Ferguson, 1982). Because available evidence indicates that the pheromones of immature queens are likely to be different from those of mature queens (pages 21 and 42), it would not be surprising if its means of distribution also differed.

Pheromone transfer in the air

Pheromone transfer by antennal contact is not the only means of signalling the presence of a queen within the nest (Fig. 2.2). Within about 30 minutes from the time a queen has been removed from a colony, or otherwise lost, the bees become agitated and search for her around the entrance. If a hive

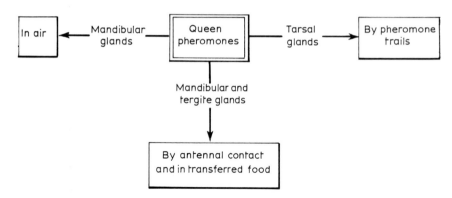

Figure 2.2 Mechanisms of transfer of queen pheromones.

containing such a queenless colony is opened many of the bees expose their Nasonov glands and fan (page 117). This behaviour could be interpreted as aiding the return of a queen to her nest, and under natural conditions might occur after the nest has been disturbed by a predator or after a queen has left to mate. The return of a queen to such a queenless colony enhances the scenting and fanning as bees move toward her.

It is not clear whether reduction in the circulation of queen pheromone by direct antennal contact would allow the queen's absence to be appreciated so quickly by this method alone, or whether another more volatile pheromone, perhaps the same as that operating within a broodless, combless cluster (page 25) also normally signals the presence of the queen within the nest. If so, perhaps different categories of workers respond more readily to the different stimuli. Certainly, workers of a queenless colony immediately perceive the return of their queen.

Queen trail pheromone

An object that has been in contact with a queen becomes attractive to workers. Butler (1969) found that a cage occupied by a queen does not lose its attractiveness completely until 90 minutes after the queen had been removed. Juska (1978) monitored the attractiveness of a cage, the mesh of which was covered with beeswax, suspended in the brood chamber of a colony. The number of bees clustered on the cage diminished to about half, 20 minutes after the queen had been removed, and to about one-fifth after a further 20 minutes. Hence, queen pheromone deposited on the wax was sufficiently persistent to suggest that the widespread trails made by the queen on the surfaces of combs form an additional means by which the queen's presence is communicated within the hive (see page 37); this would explain why bees of a queen's court that have lost contact with her often examine with their antennae the comb where she has been (Butler, 1973). The rapidity of loss of attractiveness of the beeswax-coated cage after the queen's removal is also consistent with the time taken for a colony to exhibit queenlessness.

A likely source of the queen's trail pheromone are the tarsal glands which are located on the fifth tarsomere of each of her legs (Lensky and Slabezki, 1981). Each gland consists of a unicellular layer which surrounds and secretes into a sac-like cavity. The rate of secretion of these glands of a queen is thirteen times greater than those of workers. As the queen walks over the comb her foot pads deposit the clear oily secretion onto the comb surface.

Behavioural bioassays

Various bioassays have been developed to determine the pheromones responsible for the attractiveness of queens to their workers, and for court formation within the hive. None is entirely satisfactory. A live queen may continuously produce and emit some of her pheromone components; this is difficult to duplicate in a bioassay, especially when the material concerned is in very small quantities and is highly volatile. Furthermore, it must always be borne in mind that different bioassays may be sensitive to different behavioural responses. Nevertheless, much useful and interesting, if only suggestive, information has been obtained.

One test, initiated by Gary (1961a) and developed by Butler and Simpson (1965), compares the attractiveness of cages containing different natural or synthetic pheromone components. Small cages containing different test materials are arranged on the top of the combs of a colony and the number of bees clustering on each cage is recorded after a few minutes. Either a queenless or queenright colony can be used, and the cages can be placed over combs containing brood or stores. The hive roof is supported well above the cages so they are in a dark chamber free from extraneous air currents.

Results from this bioassay reflect the ability of the test materials to attract bees to their immediate vicinity. Usually the cages are double-walled so the workers cannot touch the contents; when this is so the test does not necessarily reflect the ability of the test material to retain bees that could make contact with it, or to release antennal contact and licking. Simpson (1979) tried to overcome this difficulty by substituting the cages with small discs of blotting paper, treated with the material under test and arranged in a randomized Latin Square. Up to eight materials can be compared simultaneously. The bees can actually touch the discs and test materials and at intervals counts are made of bees whose heads are orientated toward them.

Bioassays that depend on worker bees being attracted from some distance toward the queen do not necessarily reflect behaviour in the colony where little or no long-distance movement toward the queen usually occurs. Neither do they seem to represent court formation, as the bees comprising a court have not usually moved more than a few millimetres toward their queen, but rather turn to face her when she comes into their immediate vicinity. To measure attractiveness over a distance a simple two-choice olfactometer would seem suitable (Free and Butler, 1955) although little use has been made of it.

Attempts have been made to emulate natural court behaviour more exactly by using queen models of balsa wood, pith, filter paper, plaster of Paris or polyethylene treated with glandular extracts or synthetic pheromone components.

Pain (1954a, 1956, 1961a) showed that filter paper that had been wrapped round a queen's body, and filter paper or elder pith models impregnated with extracts (in ether, chloroform, acetone or ethanol) from a queen's body surface elicited the characteristic court behaviour by worker bees.

Although worker bees may surround these models, palpate and lick them (for example, Lavie and Pain, 1959; Pain, 1961a; Gary, 1961b; Jaycox, 1970a) this does not indicate with certainty that they regard the models as queens, but that they may merely be responding to strange, attractive objects. Perhaps one way to discover whether they are treating the models as queens would be to observe their behaviour after they leave the court (see pages 10–13).

Queen mandibular gland pheromone

De Groot and Voogd (1954) found that queen pheromone was particularly abundant in the queen's head and, following the suggestion of Simpson (1956), it has been shown that the contents of the queen's mandibular glands (Fig. 2.3) inhibit both queen rearing (page 34) and development of workers'

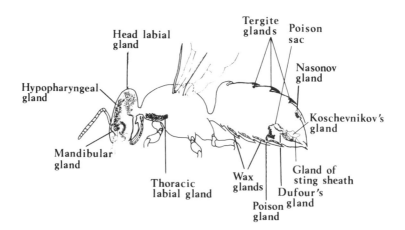

Figure 2.3 Diagram of bee indicating location of pheromone-producing glands.

ovaries (page 49). Pheromones from the mandibular glands are probably important in eliciting court behaviour.

Butler and Simpson (1965) reported that bees of a recently dequeened colony were strongly attracted to small cages containing the odour of macerated queens' heads and that this attraction was almost entirely due to the odour of the mandibular glands. The bees were not at all attracted to the odour of macerated queens' thoraces, and only very slightly to the odour of macerated queens' abdomens. The latter result is unexpected (because bees in a court pay particular attention to a queen's abdomen) and may be because the experimental design was too insensitive.

Butler and Simpson (1958) suggested that the mandibular gland secretion of the queen may be distributed over her body surface as she cleans herself, which happens two or three times an hour. However, it is possible that grooming is not essential for the distribution of pheromone on the queen's body surface; Butler et al. (1974) applied small amounts of radioactively labelled E-9-oxo-2-decenoic acid, the main component of the queen's mandibular glands (page 21), to the dorsal surfaces of the thoraces of anaesthetized queens and found that it was soon present on their heads and abdomens.

A filter paper 'queen model' that has been impregnated with mandibular gland secretion is surrounded by workers that lick it intensively (Gary, 1961b). Attempts to produce 'false' queens by coating dead (Barker, 1960) or live (Gary, 1961b) workers with the mandibular gland contents of queens has not been successful, as the normal workers of the colonies to which these 'false' queens were reintroduced were antagonistic toward them and attempted to sting them, sometimes successfully. The behaviour was similar irrespective of whether the mandibular glands came from a virgin or mated queen, or from queens of the same or a different colony to that of the recipient bees. A similar aggressive response is also often elicited when crushed mandibular glands or their extracts are smeared onto polyethylene blocks (Free et al., 1987a).

Possibly the preparations used contain 'foreign' materials that are responsible for the aggression: possibly too aggression was enhanced by release of alarm pheromone (pages 137, 145) by the examining workers; but it is more likely that the aggression may be released by an abnormal quantity or concentration of pheromone components that are attractive at the normal rate of external secretion. Further experiments, using graded doses of mandibular gland pheromone, are needed.

Queen honeybees are occasionally closely surrounded, mauled and even stung by their own workers (beekeeepers call this behaviour 'balling'), especially following a disturbance to their colony. Gary (1961b) suggested that an abnormal accumulation, followed by the release of above normal quantities, of mandibular gland secretion, might initiate such attacks. Further studies might help to elucidate this behaviour.

Perhaps the complete mandibular gland material may only attract when in association with other queen pheromones. It would be interesting to observe workers' response when it is painted onto a live queen. At Rothamsted a series of fractions of mandibular gland extracts were applied to polyethylene blocks and introduced to colonies; one of these fractions attracted a court similar to that surrounding a queen but the components concerned have not been identified chemically.

Studies of the mandibular glands' function were enhanced when Gary (1961c) devised a technique whereby the mandibular glands are extirpated from living queens which are then returned to their colonies. Such queens lose about 85% of their attraction to workers while they remain caged (Gary, 1961c; Morse and Gary, 1963; Zmarlicki and Morse, 1964; Butler, 1973). However, workers are still attracted to such a queen when she remains uncaged in her colony so the workers can touch her (Velthuis, 1970a; Butler et al., 1973), and they form an apparently normal court round her. Perhaps this is because the trace of mandibular gland pheromone that these queens contain (page 50) is sufficient to release court formation, or it could be because additional substances produced in other parts of the queen are involved. The latter seems more likely because colonies have been headed by queens without mandibular glands for 11 months and during this time they produced normal amounts of brood. Furthermore, when the mandibular glands are extirpated before they have produced any secretion and the queens are artificially inseminated normal courts are still formed (Velthuis and van Es, 1964).

These observations indicate that communication of the queen's presence does not depend solely on the secretion from her mandibular glands and provide support for the existence of another important source of queen pheromone. One such source appears to be in the abdomen itself.

Other sources of queen pheromones

Bees in a queen's court show special interest in her abdomen and an important pheromone is secreted by the tergite glands (Fig. 2.3) which are large subepidermal complexes of glandular cells located near the rear of tergites 3, 4 and 5 (Renner and Baumann, 1964). Vierling and Renner (1977) concluded that whereas the queen's mandibular gland secretion attracts workers at distances of several centimetres, the secretion from the tergite glands is only effective if workers make direct contact with the pheromone source and it has the function of stabilizing the court once it is formed. Workers of a queen's court are especially inclined to palpate with their antennae at the bases of her tergites 3, 4 and 5 (Free et al., 1987a).

Elegant experiments (Velthuis, 1967) have confirmed the attractiveness of the tergite gland secretion in the absence of the mandibular gland pheromone.

Queens whose mandibular glands had been removed within three days of emergence were used. When the ventral part of such a queen was covered with nail varnish she still elicited fairly regular court behaviour but when the dorsal part was covered (so that the tergite glands were sealed), there was only transient court behaviour. A dead worker with a queen's dorsal tergites 3 to 6 glued to its abdomen was effective in inducing court formation. The tergite glands also appear important in inhibiting ovary development (page 50).

The Koschevnikov gland (Fig. 2.3) may also be a source of attraction to court bees (Butler and Simpson, 1965). This gland is a tiny cluster of cells located in the sting chamber of the queen; its main ducts open between the overlapping spiracle and quadrate plates. When excised Koschevnikov glands were offered in double-walled cages above the brood combs of colonies they had a strong attraction to worker bees, even when in the presence of excised mandibular glands. Any attraction to Koschevnikov glands of virgin queens was much less. I am not aware that these results have been confirmed by other workers. Velthuis (1970b) was unable to demonstrate that the Koschevnikov glands inhibited ovary development of worker bees.

Chemical identification of queen pheromones

The major chemical component of the queen's mandibular glands, (E)-9-oxo-2-decenoic acid (to be referred to as 9-ODA), was identified independently by Callow and Johnston (1960) and Barbier and Lederer (1960). The amount of 9-ODA present varies with the age of the queen. Only traces (about 5 μg) are present in the mandibular glands of a newly emerged queen and about 150–200 μg in a mated laying queen with about 0.25 μg on the surface of the abdomen (Butler and Paton, 1962; Pain et al., 1967; Shearer et al., 1970).

The mandibular glands do not secrete material to the outside of the queen's body until three days after the queen's emergence (Nedel, 1960). Pain and Roger (1967) measured the amount of 9-ODA in virgin queens of 1–50 days old and discovered that those 16–20 days old had most. The heads of A. cerana and A. dorsata queens contain about 200 and 300 μg 9-ODA respectively. A. florea queens also contain 9-ODA but the quantity is unknown (Shearer et al., 1970; Sannasi et al., 1971).

Worker honeybees possess olfactory cells (sensilla placodae) on their antennae that specifically react to odours of queens and to 9-ODA (Beetsma and Schoonhoven, 1966; Kaissling and Renner, 1968).

An additional 13 components from the crushed heads of queens were identified by Callow et al. (1964) as follows: methyl 9-oxodecanoate, methyl 9-oxo-2-decenoate, methyl 9-hydroxydecanoate, methyl 9-hydroxy-2-decenoate, methyl p-hydroxybenzoate, nonoic acid, decanoic acid, 2-decenoic acid, 9-oxodecanoic acid, 9-hydroxydecanoic acid, 9-hydroxy-

2-decenoic acid, 10-hydroxy-2-decenoic acid and *p*-methoxybenzoic acid. The presence of 18 other substances was indicated but they were not identified. Only five of the latter were present in major quantities. Callow *et al.* (1964) supposed that because methyl *p*-hydroxybenzoate (also identified from queens by Barbier *et al.*, 1960) occurs naturally in many living tissues, a specific role in the honeybee would appear unlikely.

Synthetic samples of the 14 substances identified by Callow and Johnston (1960) and Callow *et al.* (1964) were bioassayed by Butler and Fairey (1964) for stabilizing swarm clusters (page 25) and attracting drones (page 94) but only 9-hydroxy-2-decenoic acid (to be referred to as 9-HDA) and 9-ODA were effective.

Later Callow (unpublished; see Simpson, 1979) identified several more substances including nonanoic acid, dodecanoic acid, tetradecanoic acid, hexadecanoic acid, octadecanoic acid, octanedioic acid, decanedioic acid, undecanedioic acid and dodecanedioic acid. To my knowledge these have not been tested.

Probably all or most of these substances identified by Callow are secreted by the mandibular glands. Some may be precursors of pheromones and others may play the role of 'keeper' substances and help to delay volatilization of the active pheromone components.

Some of the major components have been quantified in queens of different ages (Crewe, 1982; Table 2.1). The predominant component of the mandibular glands of virgin queens is (E)-10-hydroxy-2-decenoic acid (10-HDA) which is the dominant component of the mandibular glands of worker bees. As the queen ages, irrespective of whether or not she is mated, 10HDA becomes relatively less abundant and 9-ODA and 9-HDA come to predominate.

There were consistent differences between queens of the three races investigated in the relative proportions of the major components produced. The largest proportion of 9-ODA was produced by *Apis mellifera capensis* queens; *A.m. capensis* laying workers also produce a high proportion of 9-ODA (see Table 5.1).

Because the proportional blend of components produced from the mandibular gland can vary with the race of bee concerned and probably to some extent with individual queens (Crewe, 1982), the signals that the workers of each race recognize and respond to are characteristically different. The response to them is probably partly innate and partly conditioned. It has yet to be determined whether the signals produced by the queen of a race other than their own would be as effective.

The effectiveness of components other than 9-ODA and 9-HDA, singly or in mixtures, in attracting workers (see below) inhibiting reproduction (page 34) and stimulating worker activity (page 67) has not been investigated. There is no information on the chemical identity of pheromone secreted by the tergite glands.

Table 2.1 Mandibular gland components in virgin and mated queens of different races (from Crewe 1982).

Race	Condition	Number of individuals examined	Amount (μg) per head	Mean (%) present				
				10-HDA	10-HDAA	9-HDA	9-ODA	8-HOA
Apis mellifera mellifera	Virgin							
	1 day old	2	121	64	2	6	26	1
	10 days old	3	187	40	1	22	29	7
	23 days old	1	74	11	0	28	48	12
	Mated laying	8	—	15	5	32	37	8
Apis mellifera scutellata	Virgin							
	1 day old	1	50	64	7	7	12	6
	2 days old	1	21	54	4	3	33	5
	40 days old	1	927	12	2	23	49	11
	Mated laying	15	—	8	4	14	65	5
Apis mellifera capensis	Mated laying	6	—	1	1	10	85	3

Abbreviations:
10-HDA = (E)-10-hydroxy-2-decenoic acid
10-HDAA = 10-hydroxydecenoic acid
9-HDA = (E)-9-hydroxy-2-decenoic acid
9-ODA = (E)-9 oxo-2-decenoic acid
8-HOA = 8-hydroxyoctanoic acid

Response of workers to (E)-9-oxo-2-decenoic acid

The ability of 9-ODA to attract workers in the hive is in some dispute, and needs further investigation.

In tests using an olfactometer the odour of dead queens or ethanol extracts of the body surfaces of dead queens attracted workers, but the odour of pheromone from the queens mandibular glands did not (Butler, 1960c). In contrast, Butler *et al.* (1973) found a positive correlation between the ability of a queen to accumulate workers when caged over the combs of a colony and the amount of 9-ODA she contained (as determined by subsequent analysis). It was also found that the attraction of bees to cages containing filter paper impregnated with 9-ODA increased with the amount of 9-ODA. However, the possible presence of other substances which increased in amount with 9-ODA could also explain these results.

Some observers have found they were unable to induce court formation around a 'queen model' treated with 9-ODA alone (e.g. Barbier and Pain, 1960; Butler *et al.*, 1973) although 9-ODA plus mandibular gland extract was effective (Pham *et al.*, 1983); others (Free *et al.*, 1987a) obtained only a partial response. In contrast, Shaposhnikova and Gavrilov (1973) reported that a court, indistinguishable from that surrounding the queen, was formed round small wooden blocks to which 9-ODA had been added. Workers remained in the court for an average of 180 seconds, while they examined the block with their antennae (22 seconds) and licked it (4 seconds). In later experiments Shaposhnikova and Gavrilov (1975) reported that applying 9-ODA to the bodies of live queens increased their attractiveness. These differences in response reported by different authors are difficult to reconcile, and emphasize the difficulties in always obtaining consistent repeatable results in pheromone studies.

Age and attractiveness of queens

The attractiveness of adult queens increases as they grow older and become mated. Gary (1961a) showed that a virgin queen a few hours old attracts workers little or not at all, whereas an aged virgin (4 weeks old or more) is as attractive as a laying queen; when the mandibular glands of the queens had been extirpated the aged virgin was more attractive than the mated queen. Similarly, Pain (1961b) concluded that virgin queens two days old or less, do not attract workers but thereafter attractiveness increases with age. She found that confining virgin queens without workers tended to retard the development of their attractiveness, perhaps because they received insufficient food.

Although Butler and Simpson (1965) failed to demonstrate any difference between the attractiveness of live mated and virgin queens they showed that the excised mandibular glands of mated laying queens, presented in cages above a colony's combs, were strongly attractive to bees of queenless colonies. Those of aged virgins were somewhat less so, and the mandibular glands of virgin queens 7 days old or less were only slightly attractive. Using the 'roundabout' bioassay (page 123) it has been shown that bees prefer to cluster around mandibular gland contents of mated queens than of virgin queens, even in the presence of synthetic 9-ODA (Free and Ferguson, 1978).

Both the attractiveness of queens to workers (page 7) and their 9-ODA content (Table 2.1) increase as they become mated and lay eggs. Crewe (1982) produced some evidence indicating that a virgin queen does not become attractive to swarming bees until 9-ODA is present in sufficient quantity. However, a firm link between the two has not been demonstrated and may not exist. Before 9-ODA becomes abundant the presence of a high proportion of 10-HDA must give the signal produced a worker-like characteristic. Relatively little 9-HDA, another major component of the mandibular gland pheromone of mated queens (Table 2.1), is found in newly emerged queens (Crewe and Velthuis, 1980; Crewe, 1982).

It seems probable that individual queens show considerable genetic differences in the amount of mandibular gland pheromone they produce, irrespective of age differences (Pain, 1956). Provided that their offspring do not have similar proportional differences in threshold response to queen pheromone it should be possible to breed selectively for queen pheromone production.

Communication in a swarm

Soon after a swarm emerges from its hive it clusters within a few metres of its parental colony. The Nasonov pheromone released by bees as they cluster (page 116) probably attracts the queen to land, but this has yet to be shown. The vigorous fanning by scenting bees must also disperse queen scent as well as Nasonov and so presumably hastens cluster formation.

Should a swarm emerge without its queen, or the queen be unable to fly, the swarm soon returns to the parent colony, but none of the swarm bees do so when a queen is present. However, if the queen is removed from a swarm soon after it has clustered the bees usually become restless within 10–15 minutes, running about on the surface of the cluster, flying off and returning until eventually the whole swarm departs and returns to its old hive (Simpson 1963).

Because of the congestion of bees in a clustered swarm and a relative lack of movement by many of them, communication of the queen's presence is unlikely to be by antennal contact or in regurgitated food (page 8), and direct perception of the queen's odour by the clustered bees seems likely. It is known that bees are able to perceive the odour of a queen and are attracted to it in an olfactometer (Butler, 1960c).

The return of a queen to a recently dequeened swarm soon stabilizes it and the bees re-form as a quiet cluster. Simpson (1963) found that a queen in a wire screen cage also stabilized the cluster, even when the cage had double walls which did not allow the workers to contact her, but which did not prevent her odour reaching them.

Simpson (1963) also found that the total contents of a queen's mandibular glands were effective at stabilizing a cluster, but presentation of synthetic 9-ODA alone to a swarm whose queen had been removed failed to do so. However, Velthuis and van Es (1964) reported that dequeened swarms were attracted to 9-ODA presented on a dead worker bee; they were not attracted by a living queen whose mandibular glands had been extirpated and no bees exposed their Nasonov glands in her presence. Morse (1963) showed that the contents of the queen's mandibular glands were attractive to an airborne queenless swarm.

It was found later (Butler and Fairey, 1964; Butler and Simpson, 1967) that 9-HDA, also from a queen's mandibular glands, caused dispersing clusters to re-form, and that although queenless bees were attracted to a source of 9-ODA they seldom clustered on it and were only induced to form a stable cluster when 9-HDA was also present. It appeared that 9-ODA and 9-HDA together were as effective as a live mated queen, or as a queen's head that had been crushed to release the pheromones; thus 9-ODA appeared to act as an attractant and 9-HDA as an arrestant.

In contrast, Morse and Boch (1971) and Boch et al. (1975) concluded that queenless swarms are attracted less to 9-ODA or 9-ODA plus 9-HDA, than to extracts of queens' heads, and that the addition of 9-HDA to 9-ODA failed to increase its attractiveness or its swarm-stabilizing ability. Adding a mixture of synthetic Nasonov pheromone to 9-ODA or to 9-HDA attracted and stabilized queenless swarms.

Experiments by Ferguson et al. (1979) helped to resolve the previous differences. They also found that unknown components from a queen's mandibular glands encouraged cluster formation and were able to obtain stable queenless clusters in response to synthetic Nasonov pheromone mixed with 9-ODA. The addition of 9-HDA to this mixture made it less potent in initiating cluster formation, but the clusters once formed sometimes grew larger. Perhaps the greater initial attraction of 9-ODA reflects its greater volatility; the greater persistence of 9-HDA could help to prolong clustering and explain why Butler and Simpson (1967) supposed that it was necessary for cluster stabilization. These apparent discrepancies highlight the deficiencies of our bioassays, in which components are not being continually produced as from a live queen.

Therefore, it seems that a volatile pheromone from the queen's mandibular glands is responsible for initiating cluster formation and for maintaining stability. However, components produced by the queen, other than 9-ODA, which are so far unidentified, contribute toward cluster initiation and/or stability; although pheromones from the queen's head seem to be of prime importance, those from other parts of the queen's body might also be involved. It is not inconceivable that the queen may secrete components in different proportions, or that the bees may respond to different components of the queen's pheromones when she is on a comb, in a cluster or airborne.

When a swarm leaves its temporary clustering site, becomes airborne and moves to the new nest site discovered by its scout bees, the worker bees that lead probably already know the destination. They may be releasing Nasonov pheromone – this has yet to be discovered – but Avitabile *et al.* (1975) found that they could successfully 'lead' a queenright swarm by using synthetic Nasonov pheromone.

Probably the queen normally flies near the front of the swarm and her presence may help to consolidate it. Certainly bees can detect the presence of their queen in an airborne swarm. Morse (1963), Simpson (1963) and Avitabile *et al.* (1975) caged queens of clustered swarms and found that when a swarm dispersed it immediately travelled up to about 120 m (and usually not more than 75 m) in a straight line toward the chosen nest site, hovered for a while without moving forward, dispersed over a greater area than hitherto, and then returned to the caged queen.

Simpson (1963) argued that the queen must usually be near the front of the swarm for the majority of the bees to detect her, and he suggested that when bees in an airborne swarm are unable to detect the queen in front they fall back to get behind her again and perhaps force her forwards.

Morse (1963) found that when an airborne swarm was returning after a futile attempt to move without its queen, he was able to lead it by holding the queen aloft. His technique was extended by Avitabile *et al.* (1975) to lead queenless swarms as they became airborne to a new clustering site 130 m away; it was also found possible to lead these swarms with caged 'foreign' queens, methanol: diethyl ether extracts of queens, or 9-ODA alone, but they did not remain as compact clusters for long and many bees returned to their original sites. Painting the thoraces and abdomens of a few randomly-selected workers with a suspension from crushed queen's heads, from their macerated mandibular glands or 9-ODA alone, as swarms were about to become airborne resulted in these queenless swarms continuing their flight to their selected nest sites.

Therefore it seems likely that in an airborne swarm the 9-ODA component of the queen's pheromones is by itself sufficient to signal the queen's presence (and perhaps is normally the only component used for this purpose), and together with Nasonov pheromone it helps maintain cohesion and guide the swarm. Colonies in Africa moving toward far-sited nest locations may have one or more clustering stops *en route*. It is not known how the clustering site is chosen and whether the bees leading the migrating colonies are responsible. There is evidence that bees leading the swarm carry more food in their honeystomachs than those at the rear (Kigatiira, 1984) but whether the same bees lead each day is not known.

Most workers fan and expose their Nasonov glands on entering the new site. Once the queen has entered the new nest the intensity of fanning and rate of entry seems to be accelerated. Probably pheromone components other than 9-ODA are important in queen recognition and response once the swarm has occupied and clustered in its new nest. Indeed, Ferguson and Free (1981) found that at the hive entrance of a well-established colony, 9-HDA released Nasonov exposure by disorientated workers but 9-ODA had no apparent effect.

Other *Apis* species

Court formation also occurs in *A. cerana* and *A. florea*; it probably occurs in *A. dorsata* too but is difficult to observe because the queen is hidden under a canopy of bees which is several individuals thick.

There are usually 6–10 workers in an *A. florea* court (Akratanakul, 1977) but little is known of their behaviour or how queen pheromone is transferred. Preliminary experiments (Free and Williams, 1979) have been made to determine the factors that attract workers of *A. florea* to their queen.

Queen and court (*Apis florea*)

Attraction occurred when she was enclosed in a small cage so that workers were unable to touch her. The odour of a cage which had recently housed a queen was by itself sufficient to elicit a response from workers, many of which exposed their Nasonov glands. Workers were similarly attracted to the odour of 9-ODA but showed no Nasonov gland exposure. When polyethylene blocks impregnated with 9-ODA were put on the comb surface, they were surrounded by a circle of workers that looked similar to those surrounding a stationary queen. However, 9-HDA failed to elicit such a response either alone or with 9-ODA (addition of 9-HDA also diminished the attractiveness of 9-ODA to clustering *A. mellifera* workers; page 26).

Queen recognition in *A. cerana* colonies has been studied in some detail (Rajashekharappa, 1979) and has produced results which indicate some aspects might be fascinatingly different from that as understood for *A. mellifera*.

With *A. cerana*, a court (initially of 5–8 bees) soon forms round a stationary queen and the bees comprising it move their antennae faster and protrude them toward the queen, usually maintaining a small gap between her body and themselves, much as do the members of an *A. mellifera* court. Nevertheless, Rajashekharappa (1979) did not think that pheromone was transferred by direct contact. He stated that he never saw any of the workers lick their queen and only occasionally saw workers touch the queen with their antennae. However, when queens and accompanying workers were placed in double-walled cages (in which a 4 mm gap separated the inner and outer cage walls) in queenless colonies, the queen's presence was not communicated outside the cages and the colonies built queen cells, whereas when the gap between the cage walls was smaller no queen cells were built. Presumably in the latter circumstances, bees outside the cages could touch those inside and until further observations have been made there is no reason to suppose that transfer of pheromone in a queen's court does not occur by antennal contact as in *A. mellifera* (page 10).

Rajashekharappa (1979) also made a series of experiments to investigate the communication of a queen's presence within a swarm cluster. As a result of these he suggested that the presence of a queen is communicated to workers in close proximity to her by her pheromones, and this is then relayed instantaneously from worker to worker in a chain of tactile signals given by the workers' front legs and perceived by their hind legs. As a result, in a cluster, in which the bees face upwards, the queen's presence is communicated only to those bees which are above her. When a queen was caged on a vertical thread a stable cluster formed on the thread above the queen, but bees that landed on the thread below the queen were restless and did not form a stable cluster. The presence of the queen could not be communicated to a part of the cluster above the queen that was separated from the rest by a 5 cm gap. Because of the novel implications of this work it should be repeated.

Only limited work has been done with *A. dorsata*. Synthetic 9-ODA has been used to stabilize queenless groups of *A. dorsata* workers (Koeniger and Koeniger, 1980). When workers did take flight they usually returned to cluster again at the 9-ODA source.

Beekeeping applications: queen introduction

Beekeepers often need to introduce new queens to their colonies. This could be because the old queen has become lost or killed (often due directly or indirectly to beekeeping operations), or has died. Replacement could also be necessary because the old queen's progeny had characteristics which the beekeeper found undesirable, or because the old queen was more than one year old and so less able than a younger queen to stimulate brood rearing and foraging (pages 61, 65) and to inhibit queen rearing and swarming (page 38).

There are two reasons why it is difficult to replace successfully the old queen with a new one. The new queen will have an alien odour (page 105) and so will be readily recognized, and may be rejected by workers of the recipient colony. For this reason beekeepers often cage the new queen within the recipient colony for several hours so her body surface has the opportunity to absorb the odour of the recipient colony (page 103) without the workers being able to attack her. Usually, this delay in the release of the queen is achieved by closing the entrance to the cage with a layer of newspaper or a plug of candy which the bees chew through before the queen is released. However, although bees may prefer their own queen to another, in favourable circumstances a strange queen may be welcomed into a queenless colony with little or no antagonism.

The other reason why it may be difficult to introduce a new queen is because the quality and composition of pheromone she produces is different from that of her predecessor. The success obtained in introducing queen cells, virgin queens or mated laying queens to colonies is greatest when the queen of the recipient colony is either at the same stage of development or the one immediately preceeding that of the type introduced (Free and Spencer-Booth, 1961). Thus it is comparatively easy to replace a mated laying queen with another mated laying queen or a young virgin queen with another young virgin queen (Butler and Simpson, 1956). Usually no difficulty is experienced in replacing an immature queen with a virgin queen, or a virgin queen with a mated laying queen (i.e. a sequence that occurs naturally); the successful introduction of queen cells to queenless colonies increases (within limits) with the length of time the colonies have been queenless (Free and Spencer-Booth, 1961; Baribeau, 1976). In these circumstances, the introduced queen would maintain or increase the amount of queen pheromone in the colony.

When more is known about the composition of these queen pheromones it should be possible to use synthetic pheromones to maintain the status to which a colony has become conditioned. Until that time it is helpful if the walls of the cage housing the introduced queen are made of mesh wide enough for the workers to reach the queen inside with their antennae and tongues (Free and Butler, 1958) and so prevent undue reduction in the amount of queen pheromones circulating in the colony.

Conclusions

Our present state of knowledge of the functions of queen pheromone in controlling various worker activities is mostly speculative. Nevertheless, it seems that three categories of queen pheromones are involved in attracting workers.

- Queen pheromones that are present in a cluster and general hive atmosphere and whose absence enables bees to rapidly appreciate the loss of their queen must have considerable volatility.
- Pheromones attracting bees to a queen's court are probably less volatile but are still capable of causing bees a few centimetres distant to respond. Probably at a closer range they have a somewhat repellent effect.
- The queen pheromone that workers obtain on their antennae is probably of relatively low volatility. Perhaps this is the one that, as well as conveying a message of the queen's presence, inhibits queen rearing and worker ovary development (pages 36 and 48). It is especially likely to be received by bees that, because of their occupations and physiological states, would otherwise be foremost in these activities. The smaller amount conveyed by antennal contact in winter and to foragers is sufficient for the smaller needs.

Unfortunately, bioassays do not always adequately distinguish between these functions, and the sources and chemical identities of pheromones of the three categories is unknown. Presumably, the more volatile components of the mandibular glands are of importance in the first category and the less volatile pheromones in the waxy integument of the queen's body surface and from the tergite glands are of importance in the second and third categories.

Chapter three

INHIBITORY EFFECTS OF QUEENS ON QUEEN REARING

Introduction

When a queen becomes old the workers of her colony rear a new one to replace her. They also rear one or more queens when a colony reproduces by swarming; usually the old queen leaves with the swarm and one of the young queens heads the mother colony. In these circumstances the old queen lays eggs in queen cell cups (Allen, 1956) which are the empty cup-shaped precursors of queen cells and which the workers usually build near the lower edge of combs containing brood. The eggs laid in them hatch into larvae that are reared to become queens.

Queens are also reared when the old queen is missing. Under natural conditions a virgin queen may be lost on her mating flight and an old mated queen may become lost when her colony is attacked by predators. Within a few hours of the queen's disappearance the workers modify one or more cells containing eggs or young female larvae to form queen cells. This behaviour has been used extensively by research workers to study the effect of the queen on queen rearing.

Effect of worker/queen contact

Pioneer experiments by Huber (1814) demonstrated that workers need to contact their queen for queen rearing to be inhibited. Provided that a caged queen could be touched by her workers through the walls of the cage, her colony did not build queen cells. However, when the cage had double walls so that workers were unable to reach their queen inside, queen cells were built, thus demonstrating that the odour alone of the queen was insufficient to inhibit queen rearing.

More elaborate experiments by Müssbichler (1952) and Butler (1954a) confirmed and extended these results. For example, Butler (1954a) demonstrated that colonies did not build queen cells when the queen was attached to a vertical wooden board or confined to a cage with walls of queen excluder (zinc sheet with long perforations through which workers but not queens can pass), sometimes built queen cells when the queen was confined to a single-walled cage and always did so when the colony concerned was large. They always built queen cells when the queen was confined to a double-walled cage.

Müssbichler (1952) also produced evidence suggesting that only direct contact by workers of their queen inhibited queen rearing. When she divided a colony into two parts by a wire mesh screen, workers in the part without the queen built queen cells while those with the queen did not. Although this experiment demonstrated that the odour alone of the queen was insufficient to inhibit queen rearing, it is difficult to assess the amount of contact that workers on either side of the screen made with each other and so the extent to which they passed queen pheromone through it. Butler (1954a) was able to show that starving the queenless part of a colony, so encouraging food transfer (and also antennal contact) through the wire gauze screen separating parts of a colony with and without a queen also diminished the tendency of the latter part to build queen cells; however, actual lack of food in the queenless part may also have discouraged queen rearing.

Butler also demonstrated that diminishing the amount of queen pheromone workers received (either by diminishing the amount of time a queen was made available to her colony, or by increasing the number of workers in a colony), resulted in an increase in queen cell production.

For a queen to inhibit queen rearing it is therefore necessary for her workers to be able to approach her closely and make extensive physical contact with her (page 7). Hence it is somewhat surprising that tethering a queen to a comb by a short wire leash diminished her effectiveness, and queen cells were built in 17 of 42 colonies whose queens were so tethered (Butler, 1957a). Perhaps this is because under such stressful conditions the queen neither readily releases the pheromone nor distributes it over her abdomen – usually her movements over the comb deposit a pheromone trail over its wax surface (page 16) and encourage frequent changes in the composition of her court. It seems paradoxical that attaching a queen to a wooden board (Butler, 1954a) did not seem to diminish her inhibitory effect (see above).

Amputation, or binding together of the front or rear legs of a queen also induces queen rearing (Simpson, 1960; Morse et al., 1963). This could also be due to her decreased mobility and diminished ability to distribute mandibular gland secretion over her body when grooming.

Because for queen rearing to be inhibited workers need to make direct contact with their queen, it is probable that the pheromone concerned (which is relatively involatile) is both perceived and transported by the workers' antennae (page 10), producing appropriate hormonal adjustments in the workers concerned. The pheromone could come from the queen's mandibular or tergite glands or both sources.

Bioassays have been made to test the effectiveness of the mandibular gland contents, including 9-ODA and 9-HDA (see below), but not the contents of the tergite glands on queen rearing.

Queen rearing bioassays

Butler and Gibbons (1958) devised a bioassay which was used with slight modifications in subsequent tests. Groups of 50–200 worker bees were caged without queens for about 5 hours, and then each given a small piece of comb containing one-day-old larvae. When the worker bees remained queenless they constructed emergency queen cells round one or more of the larvae within 24 hours. When they were inhibited by queen pheromone few or no queen cells were produced. This inhibition occurred when an ethanol extract of mated laying queens was either mixed with sugar syrup and painted onto the bodies of dead workers attached to the comb or provided in distilled water.

Although workers usually perceive pheromone with their antennae the success of the latter method indicates that perception by taste while imbibing water or food containing the pheromone may also be effective. This needs investigation. van Erp (1960) obtained some inhibition of queen rearing by providing caged bees with the acetone extracts of one to four queens in drinking water, but obtained complete inhibition from the odour alone of a 'highly concentrated extract of queens'.

(E)-9-oxo-2-decenoic acid effect on queen rearing

Butler and Simpson (1958) showed that the contents of a queen's mandibular glands (page 19) smeared onto the body of an extracted queen, efficiently inhibited queen rearing. (For practical convenience crushed queens' heads rather than glands are sometimes used.) Although synthetic 9-ODA alone (a major component of the queen mandibular glands, see page 21) strongly diminishes the tendency of small groups of caged workers to build queen cells, it is less effective, even in large doses (e.g. 10 mg on the body of a fully extracted queen), than a recently killed mated laying queen or ethanol extracts of a mated laying queen. However, when 9-ODA is accompanied by the odour of a live mated laying queen it completely inhibits queen rearing. The odour alone of 9–ODA appears ineffective even when abnormally large quantities are present (Butler *et al.*, 1961).

Butler (1961) demonstrated that the scent from a live queen whose mandibular glands had been removed five days previously (Gary and Morse, 1960; Gary, 1961c) and so was thought to be free of mandibular gland pheromone was no less effective at inhibiting queen rearing in cages than a normal mated laying queen. Experiments with virgin queens indicated this scent was found on all parts of a queen's body; Butler (1961) concluded this queen scent alone is capable of inhibiting queen rearing in cages and that the mandibular glands produce a separate scent that is attractive to worker bees. It must be questionable as to whether the odour of a queen could be effective within the colony itself, where (as discussed previously) it is proved that workers need to contact the queen for inhibition of queen rearing to occur.

Butler and Callow (1968) supposed that the queen's inhibitory scent was largely, if not completely, equated with the odour of 9-HDA (page 22) which comes from the mandibular glands. They found that while neither the odour of 9-ODA alone nor 9-HDA alone had much of an inhibitory effect on queen rearing in cages, when presented together they were as, or almost as, effective as a mated laying queen.

Although later investigations (Butler et al., 1973) showed that even ten days after a queen's mandibular glands had been removed microgramme quantities of 9-ODA or 9-HDA or both could sometimes be detected in her head most had been removed and it is unlikely that the trace remaining could be secreted to the exterior of the queen's body. Such queens still had the ability to inhibit, at least to some extent, small and large colonies from rearing queens (Gary and Morse, 1962; Morse and Gary, 1963; Velthuis, 1970a) – again indicating the presence of another inhibitory pheromone.

Providing synthetic (E)-9-oxo-2-decenoic acid

A few experiments have been made to determine the effect of providing synthetic pheromone on queen rearing in colonies but surprisingly none have been reported in which the contents of the queen's mandibular glands have been used.

Although the mandibular glands of a queen contain 150–200 μg 9-ODA (page 21) the duration of time over which this is made available to a colony is not known. Therefore it is difficult to know how much synthetic 9-ODA simulates that provided naturally by a queen.

Chaudhry and Johansen (1971) provided the abnormally large amounts of 1000 μg, 2000 μg and 3000 μg of 9-ODA per day to medium size queenless colonies, each occupying about nine combs. Queen cells were built by all nine colonies given no 9-ODA, five of the colonies given 1000 μg 9-ODA, none of the nine colonies given 2000 μg 9-ODA and none of three colonies given 3000 μg 9-ODA. Thus, although colonies given even 1000 μg per day received more than five times the average content of a queen's mandibular glands, inhibition was incomplete. Shaposhnikova and Gavrilov (1974) found that providing queenless colonies with only 200 μg 9-ODA per day reduced queen rearing by 58%.

Doolittle *et al.* (1970) and Boch and Lensky (1976) reported that daily applications of 100 μg of synthetic 9-ODA to a porous polyethylene block suspended between the brood combs of small queenless colonies also reduced the number of queen cells built to about half that of untreated queenless colonies, but a caged queen supressed queen rearing almost completely. 9-HDA was ineffective (Table 3.1). During the swarming season in Israel Boch and Lensky (1976) applied daily doses of 2000 μg of 9-ODA (dissolved in ethanol and allowed to evaporate from a cotton wick) to medium-size queenright colonies but failed to influence the number of queen cell cups or queen cells constructed. Possibly in the last experiment the 9-ODA was presented in a manner unacceptable to the bees: in previous experiments it had been provided on polyethelene blocks (Doolittle *et al.*, 1970; Boch and Lensky, 1976) or bamboo cylinders (Chaudhry and Johansen, 1971).

The success that has been obtained using 9-ODA on small groups of bees in cages might merely reflect the relatively enormous amounts used, which were able to compensate at least partially for the absence of more appropriate pheromone components.

There is therefore a strong indication that if 9-ODA is important in inhibiting queen rearing in normal circumstances it only reaches full potential when associated with other pheromone components. Some of these may well be very volatile (Butler, 1961; Pain, 1961a) but their source and identity is unknown. The possible role of the tergite glands (page 20) in this context needs investigating.

It has yet to be proved that the mandibular gland secretion is effective at preventing queen rearing in a normal colony situation; some components in the mandibular gland secretion may inhibit queen rearing and others may attract workers to the queen to collect it. Indeed, part of the failure of 9-ODA to inhibit queen rearing completely, even in high concentrations, may be because it is not particularly attractive to workers (page 24).

Table 3.1   Effect of synthetic and natural queen pheromones on queen rearing (Boch and Lensky, 1976).

		Number of queen cells built after five days in colonies provided with:				
Experiment	*Caged queen*	*Queenless*	*Queenless + 9-ODA*	*Queenless + 9-ODA and 9-HDA*	*Queenless + 9-HDA*	*Queenless + queen extract*
1	0.2	9.5	4.3	4.1	–	–
2	–	4.4	3.0	2.7	4.4	–
3 and 4	0	3.7	1.2	–	–	0.6

It is interesting to speculate that even when bees in a colony are unable to touch their queen and obtain inhibitory pheromone, the more volatile pheromone components she produces may make them aware of her presence, and in aspects other than queen rearing the workers continue to behave as though queenright.

Age and inhibitory effect of queens

Butler (1960b) showed that ethanol extracts of virgin queens were less effective than extracts of mated laying queens at inhibiting queen rearing in cages, and that the inhibitory powers of the virgin queens increased with their age. Further cage experiments indicated that the scent of virgin queens inhibited queen rearing by small numbers of workers less effectively than did the scent of mated laying queens (Butler, 1961).

From experiments using small colonies, Butler (1957a) concluded that virgin queens are not always capable of inhibiting the workers of their colonies from rearing queens. In contrast, it has recently been demonstrated (Free et al., 1984a) that virgin queens are as effective as mated laying queens at inhibiting queen rearing in colonies of 5000 bees (Table 3.2). Experiments are needed to determine whether this is also true for large colonies. Because beekeepers can produce virgin queens much quicker and cheaper than mated laying queens they might be used as a supplementary source of queen pheromone. If one or more virgin queens could be successfully introduced and maintained in a colony already possessing a mated laying queen, the amount of queen pheromone present should be increased and so queen rearing and swarming diminished.

Anaesthetizing virgin queens with carbon dioxide encourages them to lay eggs (Mackenson, 1947). Another effect is to decrease the amount of 9-ODA they produce (Pain et al., 1967), but despite this they become more attractive to workers than untreated virgin queens, presumably because production of other attractive components has been increased (Jaycox and Guynn, 1973). If this is confirmed, comparison of chromatographs of extractions from anaesthetized and untreated virgin queens may help to indicate which peaks are important to investigate.

Influence of queen trail pheromone

The secretion from the queen's tarsal glands (page 17) may contribute toward the inhibition of queen rearing. Lensky and Slabezki (1981) found that applying the secretions from both tarsal glands and mandibular glands to combs (one queen equivalent of each per comb) inhibited queen cell cup construction, but neither was effective in doing so on its own. They observed that in an overcrowded observation hive the queen was rarely within 4 cm of

Table 3.2 Inhibition of queen rearing by immature and mature queens (after Free *et al.*, 1984a).

Colony treatment	Percentage of colonies that built		Number built per colony	
	queen cell cups	queen cells	queen cell cups	queen cells
No queen, no queen cells	100	100	5.7	4.6
No queen, open queen cells	92	67	3.0	3.6
No queen, closed queen cells	100	62	3.6	3.3
With queen	82	0	0.9	0

the bottom edge of the combs where queen cells and queen cell cups are normally constructed, whereas in an uncrowded colony she was frequently in this area. They supposed that this was probably because the clusters of bees in a crowded colony were so tightly packed at the base of the combs that the queen was reluctant to penetrate them. The authors argued that in the absence of the queen, and consequently the tarsal gland secretions she deposits at these locations, cell construction was not inhibited.

However, crowding is only one cause of swarming. In uncrowded colonies that are preparing to supersede their queens or to swarm, queen cell cups and queen cells are also constructed near the bases of the combs, and presumably queens freely walk in these areas and deposit tarsal gland secretion. Furthermore, queens often fail to make more than cursory visits to the top and side edges of combs but queen cells are not constructed in these locations.

Although it seems unlikely that the effect of the tarsal gland secretion is as localized as Lensky and Slabezki (1981) supposed, its function is well worth further investigation.

Supersedure and swarming

Colonies sometimes rear queens while their old queens are still present. When the queen of a colony that is rearing additional queens is replaced with a vigorous young mated queen further production of queen cells is usually curtailed. Similarly, when the number of workers in a queenright colony that is building queen cells is greatly reduced, the existing queen being superseded is often able to inhibit further queen cell production (Butler, 1954a,b, 1957a). Therefore, it would appear that queen cell production by a colony that is about to supersede its queen or to swarm is initiated by a deficiency in the amount of queen pheromone available to the worker bees (Fig. 3.1). This could be for one or more of the following reasons: (1) the queen is providing

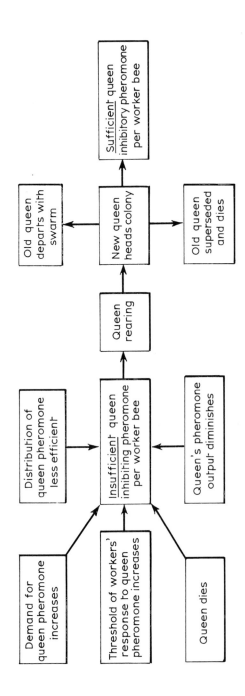

Figure 3.1 Relationship between amount of queen inhibitory pheromone and queen rearing.

less inhibitory pheromone than previously, (2) the distribution of inhibitory pheromone from the queen is less efficient than before, (3) the worker population has increased so more inhibitory pheromone is needed, (4) the threshold of the individual worker's response to the inhibitory pheromone has increased.

There is evidence (Butler, 1960a) that queens being superseded are much less effective at inhibiting queen rearing by small groups of workers than queens of colonies without queen cells. The same appears true for queens heading swarms that come from uncrowded colonies.

When a colony swarms, the old queen usually leaves with about half the worker bees to establish a new colony elsewhere. Presumably the amount of queen pheromone she produces is able to satisfy the queen pheromone requirements of the smaller population. Until it is established in a new location the swarm is obviously more at risk than the part of the colony that remains behind at the old location; it is therefore fitting that the old queen who is past her prime in pheromone production and is more expendable than a young queen, should go with the swarm. A few weeks later in the season as her colony grows in its new location, the old queen is often superseded.

Both the amount of pheromone produced per queen and the amount of queen pheromone required per worker to inhibit queen rearing are probably genetically determined and could explain why some strains of bees are more inclined to rear queens and swarm than others. Free (1961a) caged bees in their hive for the first week of their lives and found they were subsequently less inclined to depart with a swarm than bees that had always been at liberty; perhaps this was because they had become less conditioned to receiving queen pheromone and their queen pheromone requirements were less.

Based on the rate of contact and dispersal of queen pheromone by court bees (page 10) Baird and Seeley (1983) calculated that as the nurse bee population increases there is a sudden disproportionate decrease in the percentage that are inhibited from queen rearing, which could help account for the sudden onset of queen rearing following colony expansion in spring.

Lack of space in the nest may force a colony to swarm; provided that space is adequate the factors deciding whether a colony swarms or supersedes its queen are not properly understood, although they appear to be related to the time of year, environmental conditions and colony size. Perhaps these factors produce their own signals either directly, or indirectly through the queen.

When colonies swarm because they have outgrown their nests, their queens do not have diminished powers of inhibiting queen rearing (Butler 1960a) or diminished production of 9-ODA (Seeley and Fell, 1981). Nevertheless, such colonies probably also experience queen pheromone deficiency, not because the queens produce insufficient pheromone, but because during preparations for swarming the brood area of a colony becomes crowded and hinders efficient distribution of queen pheromone. Furthermore, during swarm preparation the workers harrass the queen, making her move about the comb more and reducing court formation (see page 7). The presence of queen cells in a colony possibly changes the behaviour of workers toward their queen.

Annual cycle of queen cell building

Nearly all colonies possess queen cell cups throughout the period of the year when queen rearing is likely to occur, but only a small proportion are ever used to rear queens. In fact the formation of queen cups appears to be a normal part of colony development and does not necessarily indicate that any occupied queen cells will be present subsequently (Simpson, 1959; Allen, 1965b; Caron, 1979). Perhaps workers need to mark them with a pheromone before the queen will lay in them.

Allen (1965b) found that in Scotland yearly averages of 90–100% of colonies produced queen cell cups, 5–70% had queen cells containing larvae, and 0–28% produced closed queen cells. Colonies began to produce queen cell cups in the beginning of May, and reached a yearly maximum (4–24 per colony) in late June and early July. Queen cell cups were first converted into queen cells for rearing queen larvae in late May and early June and the numbers converted per colony reached a yearly peak (from 0 to 6) from late June to the third week in July. Closed queen cells (means of only 0–1 per colony), containing pupae or spinning larvae, occurred at the beginning of July. Hence queen rearing was abortive in most colonies: more than twice as many colonies produced occupied queen cells as replaced their mature queens. Colonies producing the fewest queen cells, and which spent the least time queen rearing, were the least likely to produce adult queens.

Inhibitory effect of queen cells

It is probable, but not proven, that the number of queen cells a colony builds generally increases with its population and is an approximate measure of its queen pheromone deficiency.

Because only a limited number of occupied queen cells are present at a time in a colony it has been supposed that immature queens produce pheromones that might inhibit queenless colonies from producing additional queen cells (Butler, 1957a; Jay, 1968); this has been demonstrated only recently. Boch (1979) found that providing small, queenless colonies each with a closed queen cell greatly diminished production of further queen cells compared to queenless control colonies, but providing them with an open queen cell containing a larva appeared ineffectual. Later experiments (Free et al., 1984a) in which five occupied queen cells were introduced to each queenless colony demonstrated that open queen cells as well as closed queen cells can inhibit additional queen rearing, but not as effectively as a mated laying queen. This feedback mechanism, whereby occupied queen cells can compensate for a deficiency in inhibitory pheromone, can explain why a colony that loses its queen rears only a very limited number of possible replacements.

The source and identity of the inhibitory pheromones produced by immature queens awaits discovery. When 9-ODA is added to the wax of artificial queen cups it reduces their acceptance by the recipient colony (Ebadi and Gary, 1980). However, only traces of 9-ODA are present in queen larvae and pupae (Boch, 1979) and in newly-emerged queens (Butler and Paton, 1962). Perhaps other unidentified chemical components which inhibit queen rearing are common to both laying queens and immature stages.

Because queens are reared in the presence of mated laying queens, synthetic immature queen pheromones may be more readily accepted by colonies than synthetic mature queen pheromone. However, information so far available indicates it would be much less effective at inhibiting queen rearing.

From Allen's (1965b) observations it appears that there is often a recurring cycle involving destruction of occupied queen cells followed by replacement with further occupied queen cells within a few days. It seems probable that in a colony that originally had a queen pheromone deficiency the additional pheromones from the queen larvae more than redress the balance and lead to a feed-back mechanism which triggers destruction of the queen cells (Fig. 3.2). In some colonies this sequence will continue until late in the season when the colony no longer has a queen pheromone deficiency. However, in other colonies with a more severe pheromone deficiency, for which queen larvae are unable to compensate, some of the queen larvae will be reared to

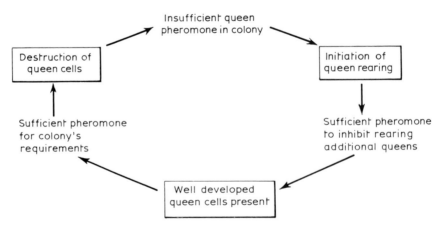

Figure 3.2 Cyclic influence of queen rearing on pheromone availability.

adults. In such colonies the presence or absence of queen cells makes little difference to the acceptance of additional ones (Free and Spencer-Booth, 1961). Colonies that swarm probably come under this latter category; they produce many queen cells (Allen, 1965b) and distribution of pheromone within them is probably hindered.

Perhaps queen pheromone requirements are increased in colonies that are preparing to swarm. The presence of only a few queen cells in colonies that are superseding their queens would seem to indicate that they are sufficient to produce the colony's pheromone requirements whereas the greater number of cells in a colony preparing to swarm still appear inadequate to meet the colony's queen pheromone requirements. A drastic division of the colony population is needed to achieve this.

Inhibiting production of queen cell cups

The construction of queen cell cups is also, at least in part, inhibited by pheromones from mated laying queens, virgin queens and immature queens, the first being the more effective (Free et al., 1984a, 1985a).

Although virgin queens inhibit queen rearing as effectively as mated laying queens they appear ineffective at preventing production of queen cell cups (Free et al., 1985a). Because queen cell cups are a necessary precursor to queen cell production their initiation is probably caused by a smaller deficit (i.e. a higher threshold level) of the same queen pheromone components that inhibit queen rearing. This would explain why virgin queens inhibit queen cell production more readily than queen cell cup production. It is also possible that queen cell cup production is normally inhibited by one or more pheromone components produced by mated queens, but which are absent or produced in smaller amounts by virgin queens. In the absence of brood few queen cell cups are built, even when only a virgin queen or no queen is present (Free et al., 1985a), indicating that the presence of brood normally helps stimulate the production of queen cell cups. Further elucidation must await chemical identification.

Lensky and Slabezki (1981) were able to induce construction of queen cell cups by experimentally crowding colonies, and so disrupting pheromone distribution, both inside and outside the normal swarming season. Above a certain threshold (2.3 workers per millilitre) the number of queen cell cups increased with colony density.

Variation in the amount of inhibitory pheromone could account for the continuous transition between shapeless knobs of wax and fully formed cups that occurs within the colony (Simpson, 1959), and for the apparent seasonal influence on cell cup construction; during the period when construction is greatest colonies are more ready to accept additional queen cell cups or to replace ones that have been destroyed (Caron, 1979). Although there is no close relationship between the numbers of queen cell cups and occupied queen cells in a colony (Caron, 1979), colonies in which queens are eventually reared might tend to begin building queen cell cups earlier than those not rearing queens (Simpson, 1959). Further observations are needed to determine how effectively the number of queen cell cups present early in the season can be used as an indication of a colony's tendency to rear queens or swarm, and whether the absence of queen cell cups can be used as a reliable indication that periodic examinations for occupied queen cells are unnecessary.

Other *Apis* species

Because the mandibular glands of *A. cerana*, *A. dorsata* and *A. florea* queens also contain 9-ODA it is likely that their secretion is normally important in inhibiting queen rearing. Butler (1966a) showed that extracts of *A. cerana* and *A. florea* queens could partially inhibit queen rearing by small groups of *A. mellifera* workers. No additional studies have been reported.

Beekeeping applications

If a synthetic product of queen pheromone that inhibits queen rearing were available commercially, its greatest value would probably be to increase the amount of queen pheromone in colonies so that no queen rearing and no swarming occurred. Absence of the risk of swarming would eliminate the need for regular routine examination of colonies to determine whether they are rearing queens, and so give enormous savings in time and labour.

Requeening could be confined to a time of year that would suit the beekeeper and risk of supersedure (known or unknown) by queens of uncertain genetic stock could be avoided. Alternatively, queen rearing could be restricted to times when supersedure but no swarming occurred.

Abrupt cessation of the supply of synthetic pheromone may facilitate replacement of the old queen with a new one, or even successful introduction of queen cells to replace the old queen by supersedure, without the need for the beekeeper to remove her (see Boch and Avitabile, 1979).

Until synthetic pheromone is available beekeepers should ensure that queens in their colonies are producing an adequate supply of natural pheromones. Because the amount of inhibitory pheromone produced depends on the age of the queen, requeening colonies at frequent intervals helps to achieve this. In commercial apiaries in southern England (Simpson, 1959) queen rearing was much more frequent in colonies with two-year old queens than in colonies with one-year old queens (70% and 11% respectively in colonies concerned). Forster (1974) concluded that in New Zealand colonies with new queens and adequate hive space rarely produce queen cells or swarm.

To minimize any deficiency in queen pheromone during the replacement of one queen by another, the new queen should be introduced directly after the old one has been removed. The mesh of the cage containing the new queen should be large enough for the bees to make antennal contact with their queen's body (Free and Butler, 1958) and the cage should be introduced into the brood area of the colony so bees which normally provide the queen's court are well placed to receive the queen pheromones.

Some beekeepers manage their honey producing colonies so that each has two mated laying queens for most of the summer. Less queen rearing and supersedure is said to occur in such colonies than in single-queen colonies (Moeller, 1976), probably because if one queen begins to fail there is still more than enough queen pheromone to inhibit queen rearing. However, earlier Farrar (1958) had concluded only that supersedure was no more frequent in two-queen colonies than single-queen colonies, and stated that when queen rearing begins in the brood nest of one of the queens it will usually follow in the brood nest of the other queen also. Clearly, more information is needed.

Chapter four

CONTROL OF WORKER OVARY DEVELOPMENT

Introduction

The ovaries of worker honeybees usually remain undeveloped, but they may show some development in the presence of a queen at certain times (Jay, 1968), and especially in the winter (Maurizio, 1954), when foraging conditions are good (Verheijen-Voogd, 1959), in colonies with poor queens (Koptev, 1957), when a colony is preparing to swarm (Perepelova, 1926) or has swarmed (Kropáčova and Haslbachová, 1970). Development is rarely enough for eggs to be laid.

However, in order that workers' ovaries may develop fully and quickly their queen must be absent. Hess (1942) found that the ovaries of 10% or more of the workers had developed to some extent in nine out of eleven colonies that had been queenless for a week; after a colony had been queenless for 13 weeks 87% of its workers had enlarged ovaries which showed all stages of transition from slight to full development. Jay (1968) observed that bees 7 days old when their colony was dequeened developed their ovaries as quickly as those less than 12 hours old; even bees 21 days old developed their ovaries although they took much longer to do so. Workers reared in a queenless colony have an increased tendency to develop their ovaries when adult (Williams and Free, 1975) which could be due to a variety of reasons, including possible deprivation of queen pheromone during larval development.

When a new queen, reared in a colony about to supersede its old queen or that has recently swarmed, is lost during her mating flight, a supply of diploid eggs and young larvae is still available in the colony from which another new queen can be reared (Fig. 4.1).

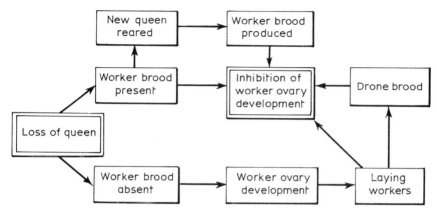

Figure 4.1 Effect of absence of queen on worker ovary development.

When the old queen of a colony dies or is killed, the workers can also rear a new queen from the worker larvae already present. However, should this new queen fail to return from her mating flight there are no young larvae available to rear a replacement. (It would be advantageous if workers protected any other virgins present and prevented them from emerging from their cells until the new queen's safe return).

A similar situation arises when a queen dies or is killed in a colony without brood present. Larvae are absent when a colony has just established itself in a new nest and during part of the winter. Perhaps the tendency of queens to lay eggs throughout the winter even though workers do not rear the brood (Free *et al.*, 1987a) is a safeguard against their queen becoming lost.

Under all such circumstances when the colony itself is doomed, up to 25% (Perepelova, 1928) of the workers develop their ovaries fully and lay unfertilized (haploid) eggs which eventually produce drones that might propagate the colony's genes. Workers that lay eggs do so for relatively short periods only and between times continue their normal duties.

Soon after a queen has been successfully introduced to a colony with laying workers their ovaries begin to degenerate and after a week or so the proportion of workers with developed ovaries is indistinguishable from normal (Milojević *et al.*, 1963; Husing and Bauer, 1968).

Inhibitory influence of queen

When Hess (1942) divided a colony by a single wire mesh screen ovary development did not occur in either the part with or without the queen; but when a double screen was used so that bees in the queenright and queenless parts could not contact each other, the ovaries of workers in the queenless part developed. Hess suggested that a substance produced by the queen is distributed in food to her workers and inhibits the development of their ovaries. While this explanation now seems doubtful her experiments demonstrated that contact with the queen, or with workers that had contacted the queen, was necessary for inhibition of worker ovary development.

Bioassays were developed by de Groot and Voogd (1954) and Pain (1954b) in which groups of 50 worker bees were confined in cages provided with a piece of comb, water and a mixture of sugar and pollen; queen pheromone was provided in some cages but not others. After three weeks the worker bees were dissected and the extent to which their ovaries had developed was determined.

The first experiments of this type demonstrated that a living mated or virgin queen, or the body of a dead mated queen, even of one that had been dead for several years, inhibit ovary development (de Groot and Voogd, 1954; Pain, 1955; Verheijen-Voogd, 1959). But for inhibition to occur the workers need to be able to touch the queen and her odour alone is insufficient; when a queen was caged among workers so that they could not touch her, inhibition was not achieved (de Groot and Voogd, 1954).

Velthuis (1972) found that dead queens were only effective when workers could make direct contact with them, and they were unable to inhibit ovary development of bees the other side of a wire screen.

Transfer of queen pheromone

Voogd (1955) found that dead queens she had extracted with acetone no longer inhibited ovary development, but did so again when impregnated with acetone extracts from other queens. Administration of queen extract could cause retrogression of workers' ovaries in a queenless colony (Milojević et al., 1963).

Voogd (1955) concluded that the queen pheromone extract needed to be applied to an object, such as a dead worker bee, before inhibition of ovary development was complete. However, Butler (1956) added bees that had recently licked their queen to a group of queenless bees and found this treatment helped to suppress their ovary development (in the light of recent developments it would be interesting to introduce instead workers that had palpated their queen but not licked her; page 11). He also found that ovary development of caged bees was suppressed by giving them the excised foreguts, including the honeystomachs, of bees of queenright colonies.

Introducing extracts of queens into the food provided for caged bees was also effective to some extent, but less so than providing it on the body of a dead extracted queen or 'model' queen (Voogd, 1956; Butler, 1957b). Pain (1961c) concluded that antennal stimulation as well as ingestion of the pheromone is needed. Workers could more readily make antennal contact with the queen's pheromone when it is presented on a dead queen's body.

Because the offering is as effective on an inanimate object as on a dead queen it now seems unnecessary to resort to a 'psychological stimulus' to help explain the results.

It is likely that the perception of pheromone by sensory cells in the antennae triggers the release of an endrocrinological response within the recipient bees which in turn inhibits ovary development. It is doubtful whether the queen's inhibitory pheromone is transmitted in the food, thence through the intestinal wall into the haemolymph and has a direct biochemical effect on the ovaries; furthermore, the pheromone would probably become too dilute in the transmitted food for this to be effective. However, the queen's pheromone may well be perceived gustatively and then act through the central nervous system.

There is also a strong indication that queen pheromone is ineffective when presented in sugar syrup of higher concentration than 5–10% (Verheijen-Voogd, 1959; van Erp, 1960; Pain, 1961c), which is well below the concentration of most nectar or honey circulating in the hive. Until more evidence is available the results obtained using techniques in which the worker bees imbibe the pheromone in syrup should not be accepted without reservation.

Source of inhibitory pheromone

De Groot and Voogd (1954) and Pain (1955) found evidence that the queen's inhibitory pheromone was particularly abundant on her head, the contents of which alone could prevent ovary development, and Butler (1959) showed that the mandibular glands in the head were responsible. In cages containing dead extracted queens treated with mandibular gland extracts only 19% of worker bees showed some ovary development, compared to 65% in cages containing dead extracted queens only.

Following the identification of (E)-9-oxo-2-decenoic acid (9-ODA) as the major component of the mandibular glands (page 21) it was found to suppress ovary development but considerably less effectively than a queen extract (Butler et al., 1961; Velthuis and van Es, 1964). Unfortunately, giving 9-ODA in food or on the body of an extracted queen precludes an exact measurement of the amount of material taken by the worker bees. However, relatively large quantities of synthetic 9-ODA were provided in the above experiments (about 50 µg per 40/50 worker bees per day). Velthuis (1972) found that it was sometimes necessary to provide 30 µg or even 150 µg to 50 queenless bees every three days before any inhibition occurred. Hence it seems that other components (which have not yet been identified) in the queen's mandibular glands must be important. A mixture of one queen equivalent of 9-ODA (150 µg) and 9-HDA (75 µg) was less effective than the extract from one queen's mandibular gland (Velthuis, 1976). Queens of

A. mellifera capensis, in whose colonies laying workers readily develop, produce proportionally more 9-ODA in their mandibular gland secretion than do queens of other races (Crewe, 1982).

Butler and Fairey (1963) injected the haemolymph of workers with 15 μg 9-ODA each and found 14 days later that despite such large doses, inhibition of ovary development was only partial; this could be either because this is not the normal mode of 9-ODA activity, or because 9-ODA alone is insufficient, or both.

Pain (1961c) decided that both 9-ODA and the odour of a queen are necessary to inhibit worker ovary development. She found that 9-ODA alone was insufficient to do so whether presented on food or on the bodies of dead workers. However, Butler and Fairey (1963) showed that although the scent alone of a live mated queen sometimes slightly diminished ovary development (confirming the role of sensory perception on inhibition), it does not act synergistically with 9-ODA and both together are much less effective than allowing workers direct access to a live mated queen.

Velthuis (1970b, 1972) reported that queens whose mandibular glands have been extirpated (page 20) can still inhibit ovary development. He pointed out that because the amount of 9-ODA on the body of such a queen does not exceed 0.25 μg (Butler and Paton, 1962), when she is introduced to a colony of 25 000 bees each will receive at most only 10^{-11} μg of 9-ODA during the whole period the queen is present. In the above experiments, amounts of 9-ODA far in excess of this failed to inhibit queenless bees from developing their ovaries. Other components must have been responsible, and it was unlikely they came from the mandibular glands. Perhaps pheromones from one or more sources other than the mandibular glands are normally important, and are able to replace one another to a large extent when necessary. Other experiments give support to this suggestion.

The queen's mandibular glands barely secrete in the first three days of her adult life and a queen 1–2 days old has only about 5 μg 9-ODA present (Shearer *et al.*, 1970). Nevertheless, when the mandibular glands are extirpated before a queen is 3 days old and she is artificially inseminated, some inhibition still occurs (Velthuis 1970b, 1976). In an intact queen in which the mandibular gland secretion is distributed over her body, the head, thorax and abdomen have approximately equal inhibitory powers but queens without mandibular glands have greater inhibitory powers in their abdomens than in their heads and thoraces.

Further experiments (Velthuis 1970b, 1976) indicated that the Koschevnikov glands are ineffective and only the dorsal side of the abdomen bearing the tergite glands (page 20) suppresses ovary development. The tergite glands of queens with mandibular glands have much greater inhibitory powers than those of queens without mandibular glands, indicating that the tergite and mandibular gland secretions normally work together, perhaps synergistically.

The apparent ability of either the tergite glands or the mandibular glands to inhibit ovary development independently is further evidence that inhibition is initiated by chemoreception followed by hormonal adjustments, rather than by biochemical means.

In many of the cage bioassays each bee probably makes frequent and direct contact with the pheromone source provided, and much more so than occurs in the hive or nest. Experiments with normal colonies have confirmed the important role of queen pheromone in inhibition of ovary development; they also emphasize the importance of brood pheromone.

Inhibitory effect of worker brood

Perepelova (1928) stated that the presence of abundant larvae in queenless colonies delays the appearance of laying workers. Milojević and Filipović-Moskovljević (1958, 1959) discovered that worker larvae and pupae had an inhibitory effect on worker ovary development in very small queenless colonies (350 bees), the larvae being the more effective. An inverse relationship has been found between the amount of worker ovary development and the number of worker larvae in colonies that have swarmed (Kropáčova and Haslbachová, 1970).

Jay (1968) began a series of important and intensive experiments to determine the relative importance of queen pheromones and of brood pheromones in influencing worker ovary development in colonies several thousand strong. His work demonstrates how necessary it is to follow up cage experiments with experiments on full-size colonies before their results can be properly interpreted.

It had previously been shown that virgin queens and queen pupae had little inhibiting influence on ovary development in queenless colonies of only 500 workers (Milojević et al., 1963), and Jay (1968) demonstrated that providing queenless colonies with virgin queens, queen cells containing larvae or pupae, or allowing colonies to rear their own queens failed to inhibit the development of workers' ovaries; likewise the presence of queen larvae or queen pupae in queenless and broodless colonies was ineffectual (Jay, 1970). These surprising results are in striking contrast to the marked inhibitory effect of virgin and immature queens on queen rearing (pages 37 and 41). However, although queen brood is ineffectual, the presence of worker larvae and pupae suppresses worker ovary development as strongly as a mated laying queen (Jay, 1970; Kropáčova and Haslbachová, 1970, 1971).

Experiments under field conditions (Jay, 1972) demonstrated that only a small amount of ovary development occurred in colonies with brood and a queen. Removal of the brood greatly increased ovary development whereas removal of the queen from broodless colonies only slightly increased ovary development. Thus, although the queen had some inhibitory powers (especially in the absence of brood) the brood's influence was of great importance. Larvae and pupae were equally effective. In later experiments using small colonies of about 5000 bees the queen and brood proved to have similar inhibitory powers, but the presence of both did not have a greater effect than either on its own (Jay and Jay, 1976).

Jay (1972) also divided queenless colonies by a screen (single or double) into two parts; one part contained brood, the other part did not but its bees were of course exposed through the screen to the odour of brood. Ovary development in the parts without brood was much greater than in the part with brood. However, it was less than in equivalent queenless and broodless colonies which were not subjected to the odour of brood. Therefore, it is apparent that nearness to or contact with brood has the greatest inhibitory effect, but the odour alone has some influence. The brood pheromone that stimulates foraging (page 75) has a similar mode of action. Unfortunately no difference was apparent from using a double or a single screen. Probably this was because bees on the two sides of the screen make little attempt to contact each other. It is difficult in this type of experiment to know the amount of contact that has occurred through the screen (page 33), and consequently the results are difficult to interpret.

The volatile scent from the brood (which has some inhibitory power) does not appear to emanate from brood by-products such as faeces, or from cast skins and cocoons left in vacated cells. Thus, a high incidence of ovary development can occur on old brood combs in queenless, broodless colonies (Jay, 1972) and there is no difference in the extent of worker ovary development in colonies transferred to new comb or to old comb (Jay and Jay, 1976); indicating that the inhibiting material is almost certainly a pheromone produced by the brood itself.

Kubišová and Haslbachová (1978) provided caged bees with ethanol or acetone extracts of worker larvae in honey, on dead worker bees or on polystyrene blocks. All methods, but especially the first, helped suppress worker ovary development. In subsequent experiments Kubišová et al. (1982) demonstrated that those fractions of dichloromethane extracts that contained acids or mixtures of acids and steroids had the greater inhibitory effect. The brood pheromone concerned could be the same as the brood recognition pheromone (pages 69–74).

Whereas brood pheromone inhibits ovary development it is probably necessary to stimulate development of the hypopharangeal glands (which produce brood food) of young bees; it seems to be necessary for the bees to make contact with the brood, the odour alone of the brood being insufficient (Free, 1961a).

Lack of inhibitory effect of (E)-9-oxo-2-decenoic acid on queens

The ovaries of a queen honeybee develop normally despite the presence of a high concentration of 9-ODA in her mandibular glands and ample brood pheromone in her colony. Pain and Barbier (1981) were unable to detect

9-ODA in extracts of queens' ovaries or haemolymph, and found that injecting 9-ODA into a queen's haemolymph failed to influence ovary development. As a result they suggested that 9-ODA is specifically deactivated in queens by transformation into metabolites or by reaction with a protein.

Other *Apis* species

Little worker ovary development occurs in *A. cerana* colonies with mated laying queens and brood (Bai and Reddy, 1975). Brood alone has an almost equivalent inhibitory effect, but in a broodless and queenless colony only 8% of the worker bees had undeveloped ovaries and 72% of them had well-developed ovaries. As with *A. mellifera* (page 47) inhibitory pheromones from the brood had the greatest influence when workers could contact it, although brood odour alone had some influence.

Extracts from *A. cerana* and *A. florea* queens have been found to inhibit the ovary development of *A. mellifera* workers (Butler, 1966a), but the influence of *A. cerana* and *A. florea* queens on workers of their own species in broodless colonies has not yet been investigated. Laying workers soon make their appearance after the removal of a queen from an *A. cerana* colony and they may continue to lay eggs after a new queen has emerged (Sakagami, 1958). Butler (1967) stated that *A. cerana* workers very readily develop their ovaries during preparations for swarming and supersedure, and some of them lay eggs.

In contrast Velthuis *et al.* (1971) concluded that very few (and possibly only one) active laying workers are present at a time in an *A. dorsata* colony; if this is so each probably produces abundant inhibitory pheromone (page 59).

Conclusions

The mandibular gland contents are capable of inhibiting ovary development of small groups of worker bees in cages. This is due to components, not yet identified, other than or in addition to 9-ODA and 9-HDA. However, super-normal quantities of the latter two components can compensate to some extent for lack of the missing components.

The tergite glands also have inhibitory powers, and are largely responsible for suppressing ovary development in queens whose mandibular glands have been extirpated, but they probably normally act in conjunction with the mandibular gland contents. The secretion from the tergite glands has yet to be identified chemically. Perhaps the queen's pheromones concerned are partially the same as those which inhibit queen rearing; this would account for some ovary development in colonies preparing to swarm.

However, in a normal colony inhibition from brood pheromone is as effective as from queen pheromone. When a colony becomes queenless the worker brood pheromone present suppresses, to a considerable extent, the development of the workers' ovaries. The colony will begin to rear queens, but while still immature the queens do not aid suppression. Within a few days of becoming adult the queen's mandibular glands secrete inhibitory pheromone; possibly the tergite glands secrete earlier still. Usually, the new queen will have mated and begun to lay eggs while some worker pupae whose inhibitory effect is as great as worker larvae are still present, and worker ovary development will be severely curtailed. However, should for some reason the worker brood all have emerged before the queen has begun to lay eggs, the queen's mandibular glands and tergite glands alone are responsible for inhibition. After the queen has begun to lay, the brood pheromone also makes an important contribution. This contribution presumably increases, to some extent at least, with the amount of brood present.

If, for some reason, the colony becomes queenless after the worker brood have all emerged so there is a complete absence of inhibitory pheromone, many of the worker bees develop their ovaries and lay eggs. Pheromones from these laying workers and their brood help restrict worker ovary development (page 56), and so collaboration and co-ordination is maintained in the doomed colony.

Beekeeping applications

To prevent production of laying workers and their brood in queenless colonies, and associated difficulties in introducing new queens, the colonies should not be allowed to become short of worker brood. If such a condition has arisen it should be possible to rectify it by introducing large amounts of worker brood with their inhibitory pheromone. When synthetic brood and queen pheromone are available they can be used for the same purpose.

Chapter five

LAYING WORKERS

Behaviour

In a honeybee colony that is broodless and queenless the ovaries of many of the workers develop to some extent, and a few workers have ovaries that are sufficiently well developed for eggs to be laid (Perepelova, 1928; Sakagami, 1959; Jay, 1968). Most of the workers whose ovaries reach full development lay eggs (about 20–30) for a short period (4–6 h) only, and also undertake normal worker duties, including foraging. However, when they are laying, other workers lick them and palpate them with their antennae (Perepelova, 1928).

Sometimes in a queenless colony one of the workers is constantly surrounded by a 'court' of attendants and appears to be treated as a queen (Park, 1949; Lunie, 1954). In each of several small queenless colonies, Sakagami (1958) observed that one of the workers attained the status of a 'false queen' and was constantly attended by 1–20 workers (usually 3–8). A false queen had the same appearance as an ordinary worker except for a slightly extended polished abdomen. She appeared to behave in all ways like a true queen, spending all her time either resting, egg laying or walking on the comb, and undertook no ordinary worker duties. Such false queens had a much higher egg-laying rate than the usual laying workers.

When a false queen moved over the comb workers gave way to her; when she was stationary they palpated her with their antennae and licked her, especially the terminal segments of her abdomen. On losing a 'false queen' the workers behaved as if they had lost a true queen, and another laying worker soon took her place. It would be interesting to determine the effectiveness of various intermediaries between worker and queen, that can be produced experimentally, as substitute queens.

Sakagami (1958) observed that on rare occasions a false queen was mauled aggresively and she responded by stropping her tongue, thereby possibly distributing an 'appeasement pheromone'. Such aggressive behaviour is common in queenless colonies. At the time egg laying began in a queenless colony Sakagami (1954) noted that the workers developed aggressive behaviour, in which many of the behaviour patterns associated with antagonism toward intruders at the hive entrance (Butler and Free, 1952) were exhibited. No definite hierarchy based on dominance and submission was established, probably reflecting the very large number of individuals involved. However, those bees that were mauled had well-developed ovaries, while the ovaries of aggressive bees showed a wide range of development. Velthuis (1976) confirmed that laying workers are subjected to aggression and respond by being submissive, but found that the ovaries of all the aggressive bees showed some development.

The different responses to laying workers and false queens are probably associated with the amount and blend of the pheromone components each type produces (page 57).

Inhibitory effect of laying workers

Sakagami (1958) noted that while a false queen was present in a colony ovary development and oviposition by other workers diminished. The introduction of a laying worker or extract of a laying worker into a cage containing queenless newly-emerged workers evokes court behaviour and has a slightly inhibitory influence on the development of their ovaries (Velthuis *et al.*, 1965; Velthuis 1970b). In a broodless colony the presence of laying workers helps to inhibit ovary development of the remainder, although not as effectively as a mated laying queen. Nevertheless, the presence of laying workers in conjunction with haploid drone brood produced by the laying workers themselves suppresses ovary development as effectively as the queen and her diploid worker brood (Jay and Nelson, 1973), and helps explain why the proportion of workers with enlarged ovaries in a colony is always limited.

In contrast to the inhibitory effect on ovary development of laying worker pheromone, there is an indication that it may stimulate comb building (Darchen, 1957).

Apis mellifera capensis

Workers of a South African race of honeybee, *A. mellifera capensis*, (the 'Cape honey bee') lay eggs that develop without fertilization into diploid females. Queenless colonies of *A. mellifera capensis* readily produce false queens which lay up to 200 eggs per day.

When an *A. mellifera capensis* worker is introduced to a cage or colony containing queenless *A. mellifera* workers she rapidly establishes herself as a false queen (Velthuis, 1976); a court is formed round her within four days and she begins laying within 7–10 days of introduction and she may continue laying for 3–5 months.

However, within a group entirely of *A. mellifera capensis* workers the development and characteristics of laying workers are not different from those of other races.

Pheromone components

Laying workers of *A. mellifera capensis* produce abundant 9-ODA (Ruttner *et al.*, 1976) but there is no direct relationship between 9-ODA production and oviposition; none was found in 9% of *A. mellifera capensis* workers that laid eggs and in 33% of those that did not (Hemmling *et al.*, 1979).

Mandibular glands of the *A. mellifera* worker produce (E)-10-hydroxy-2-decenoic acid (10-HDA) which is the main component of the brood food given to worker and queen larvae. Crew and Velthuis (1980) found that in addition to 10-HDA the mandibular glands of laying *A. mellifera* workers contain two components, 10-hydroxy decenoic acid (10-HDAA) and (E)-9-hydroxy-2-decenoic acid (9-HDA), that are produced in the mandibular gland secretion of queen honeybees; (E)-9-oxo-2-decenoic acid (9-ODA) was absent (Table 5.1). However, the mandibular glands of one exceptional laying *A. mellifera* worker also contained 8-hydroxyoctanoic acid (8-HOA) and a considerable proportion of 9-ODA. A somewhat higher incidence of 9-ODA was reported by Saiovici (1983); he found 9-ODA comprised 10% of the mandibular gland secretion of *A. mellifera* workers with developed ovaries. Velthuis and van Es (1964) presented 9-ODA on a dead worker in a cage with 50 queenless workers and found that a court, sometimes 10–15 workers strong, was formed round it.

In contrast to *A. mellifera*, 9-ODA forms about 75% of the total mandibular gland secretion of laying workers of *A. mellifera capensis* (Crew and Velthuis 1980; Saiovici, 1983) and exceeds the proportion of 9-ODA found in a queen's mandibular gland secretion. Four of the other six components found in the queen's mandibular gland extracts are also present (Table 5.1).

The presence of 9-ODA in some *A. mellifera* workers serves to illustrate that their mandibular glands are capable of producing it; perhaps 9-ODA synthesis is characteristic of those *A. mellifera* laying workers that become false queens. The ability of laying *A. mellifera* workers to inhibit further worker ovary development within their colonies cannot normally be dependent on 9-ODA production and the signals they release must be different from the queen's. This could help explain why they sometimes release the aggressive responses of other workers, whereas false queens, producing 9–ODA, release court behaviour (page 55). Probably the difficulty beekeepers find in successfully introducing queens to colonies that have been queenless for some time, reflects an adequate supply of 9-ODA in the colony.

Table 5.1 Mandibular gland components present in queens and laying workers (from Crew and Velthuis, 1980).

Category	Number of individuals examined	Amount per head (µg)	Mean % present						
			10-HDA	10-HDAA	9-HDA	9-ODA	8-HOA	4-HOB	M-HPE
A. mellifera virgin queens one day old	7	137	62	2	7	26	2	1	0
A. mellifera mated queens	5	197	12	8	32	36	7	2	3
A. mellifera workers in queenright colony*	–	–	74	16	2	0	3	–	–
A. mellifera laying workers	12	6	77	18	5	0	0	0	0
A. mellifera laying worker (exceptional)	1	22	53	8	12	18	9	0	0
A. m. capensis workers with partially developed ovaries	17	61	42	3	7	34	14	0	0
A. m. capensis laying workers	3	22	5	1	8	76	10	0	0

Abbreviations: 10-HDA = (E)-10-hydroxy-2-decenoic acid
10-HDAA = 10-hydroxy decenoic acid
9-HDA = (E)-9-hydroxy-2-decenoic acid
9-ODA = (E)-9-oxo-2-decenoic acid
8-HOA = 8-hydroxyoctanoic acid
4-HOB = methyl p-hydroxybenzoate
M-HPE = 2-methoxy-4-hydroxy phenylethanol

* From Velthuis 1985

It would be interesting to know whether laying workers could also inhibit queen rearing if young diploid brood were provided.

The ability of *A. mellifera capensis* to release queen-like signals could explain the rapidity with which they establish themselves as false queens.

It would also be interesting to discover the amount of variation in the proportions of components present that commonly occurs between individual laying queens and between individual laying workers of the same race (see page 23). If little variation occurs, the proportional blend of components could well be an important distinguishing feature. If there is much variation, then the actual quantity of one or more of the components present is probably of paramount importance in distinguishing between categories; so that even if the proportion of 9-ODA in the mandibular gland secretion were greater in a laying worker than in a queen, the queen is recognized by the much greater amount of 9-ODA she contains.

Pheromones other than those produced by the mandibular glands may be involved. Saiovici (1983) demonstrated that the removal of the mandibular glands from 2–3 day old *A. mellifera capensis* workers did not reduce their ability to inhibit the ovary development of groups of twenty *A. mellifera* workers. Observations are needed to determine whether such *A. mellifera capensis* workers are still regarded as false queens and retain the ability to elicit court behaviour. Possibly inhibitory signals from two different glands are normally produced by false queens, either of which on its own is able to compensate for the absence of the other. Similar experiments on the effect of removing the mandibular glands from *A. mellifera* laying workers and false queens should be undertaken.

Other *Apis* species

A. cerana workers develop their ovaries even in the presence of their queen. Sakagami (1954) found that 12 of 20 bees from the queen's court had fully developed ovaries and five had partially developed ovaries, indicating that the presence of developed ovaries did not influence the behaviour of workers toward their queen. In a queenless *A. cerana* colony a false queen attracted a court of 3–20 workers that licked her body, and especially the tip of her abdomen. When *A. mellifera* workers were also present they occured relatively more frequently than *A. cerana* workers in the false queen's court (Sakagami, 1958).

Little information is available about *A. dorsata* laying workers. It seems probable that the few laying workers that are present in a queenless *A. dorsata* colony may inhibit the rest from developing their ovaries (Morse and Laigo, 1969).

Primitively eusocial bees

In a colony of *Lasioglossum zephyrum*, a primitively social halicitid bee, the workers with the largest ovaries and which nudge others most often are also nudged most frequently by the queen (Breed and Gamboa, 1977). Possibly workers with developed ovaries are recognized by the pheromone they produce.

Chapter six

STIMULATORY EFFECT OF QUEEN PHEROMONE

Introduction

In addition to inhibiting reproduction by workers, either through development of their own ovaries or through queen rearing, the pheromones produced by a queen stimulate many worker activities including comb building, brood rearing, foraging and food storage (Fig. 6.1). In fact, it appears that in the absence of sufficient queen pheromone many worker bees must spend much of their time unoccupied. The production by one individual of pheromones that induce the remainder to work harder was obviously a most important step in the evolution of social life. Thus, queens of primitively eusocial bees may have a general stimulatory effect on the activity of other members of their colonies. Removing queens of the halictid bee *Lasioglossum zephyrum* from their small colonies reduces activity to about a third (Breed and Gamboa, 1977).

Brood rearing

With the same rate of queen pheromone production the amount available per worker bee must be greater in small than in large colonies. This appears to be reflected in the comparative activities of small and large colonies.

In autumn, small colonies tend to continue brood rearing later than large colonies (Farrar, 1932; Moeller, 1961; Free and Racey, 1968); during autumn and spring the amount reared per bee in colonies with brood diminishes with increase in colony size (Figs 6.2 and 6.3). Colonies headed by queens of the current year are also likely to continue brood rearing later into the autumn than colonies headed by older queens (Free and Racey, 1968),

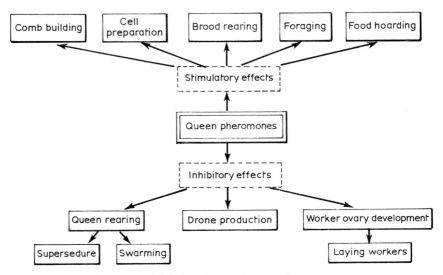

Figure 6.1 Inhibiting and stimulating effects of queen pheromones.

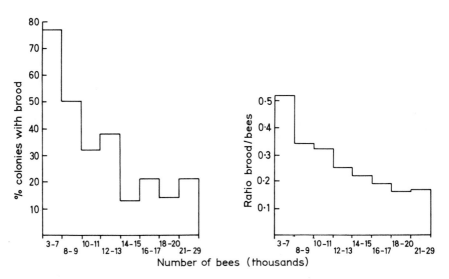

Figure 6.2 Relationship in autumn between the number of worker bees in colonies and the presence of brood (left) and between the number of worker bees in colonies with brood and the ratio of brood/bees (after Free and Racey, 1968).

and are reported to have twice the amount of brood in winter (Avitabile, 1978); current year queens have the greater pheromone production.

During the latter half of the winter in temperate climates, as the adult bee population diminishes, brood rearing increases. If a constant supply of queen pheromone were available it could be supposed that, as the winter progresses, the reduced adult population receives more queen pheromone per bee and so a greater stimulus to rear brood. However, according to Pain *et al.*, (1972), the amount of 9-ODA produced by a queen varies greatly throughout the year, reaching a major peak in June and a secondary peak in December. Such a variation could greatly influence the amount available per bee at different times of the year, and so could in turn have a profound effect on colony growth and development. Further studies are needed.

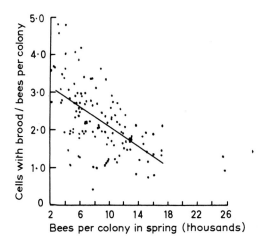

Figure 6.3 Relationship in spring between the number of worker bees in colonies and the amount of brood per bee (from Free and Racey, 1968).

Comb building

The presence of the queen greatly influences the amount of comb built. The minimum number of worker bees required to be together before comb building begins is lowest (50 bees) in the presence of a laying queen, and progressively increases in the presence of a virgin queen (75 bees) and a dead queen (200–300). Queenless bees build no new comb irrespective of the size of their population; although they may still continue to build on foundation (wax sheets embossed with the shape of cell bases) if this has been provided by the beekeeper (Darchen, 1957). In colonies headed by old or failing queens comb building is greatly curtailed or ceases.

Queen pheromone therefore appears to be of prime importance in stimulating building. The smaller the colony, the greater the stimulus each building bee is likely to receive. Taranov (1959) demonstrated that the amount of wax produced per individual young bee is inversely proportional to the number of them in a colony:

Number of young bees in colony	Production of wax (g)	
	per colony	per 1000 bees
1100	15.5	14.1
1800	19.1	10.6
2400	20.1	8.4
3000	25.1	8.3
3700	29.2	7.9
5000	33.0	6.5

Darchen (1968) divided a colony into two parts that were identical except one part contained the queen and the other was queenless. The part with the queen subsequently built four to five times as much comb as the queenless part. Such comb building as occurs in queenless colonies is probably stimulated by the presence of immature queens being reared (Free, 1967a). Taber and Owens (1970) made a thorough study of comb and cell construction in 'natural' colonies; they concluded that queenless colonies built cells, used for rearing drones and storing food, that were often irregular in shape (having side walls of varying lengths) and were intermediate in size between normal worker and drone cells. Perhaps, therefore, adequate queen pheromone enhances the quality as well as the quantity of work produced. When no brood is present in queenless colonies no cell building occurs (Free, 1967a).

Darchen (1968) demonstrated that the odour alone of a queen was inadequate to induce workers to construct comb – they needed to be able to touch her. Either the head, thorax or abdomen of a dead queen was sufficient to stimulate comb building in a small queenless colony but the head was especially effective. When a colony was divided between two compartments of a hive with the queen secured in a rubber diaphragm in such a way that her head was in one compartment and her abdomen in the other, comb building only occurred in the compartment containing her head. Comb building by small groups of worker bees in cages can be stimulated by acetone and ether extracts of queens' heads or mandibular glands (Chauvin et al., 1961). However it is necessary for the extract to be administered on paper strips so the bees can contact it; when mixed with candy before presentation it is ineffective. It would appear that the pheromone must be distributed by antennal contact (page 11).

Free et al., (1987b) confirmed that laying queens stimulate comb building and showed that virgin queens were equally effective. Colonies with only immature queens produced no more comb than queenless colonies, so it appears that immature queens are ineffective at stimulating comb building although they are effective at stimulating foraging (page 68) and inhibiting queen rearing (page 38). Possibly all three categories of queen (i.e. mated laying, virgin and immature) use the same pheromone components in stimulating foraging and inhibiting queen rearing but different ones are involved in stimulating cell building, or immature queens use different components from mature queens for inhibiting queen rearing and stimulating foraging and these are ineffective at stimulating comb building.

Darchen (1968) suggested that the presence of laying workers may stimulate comb building. In queenless colonies, containing about 5000 worker bees, building began about 14 days after the queen had been removed, simultaneously with the appearance of laying workers and eggs in the cells. It has since been shown that the mandibular glands of laying workers of the race *A. mellifera capensis* (page 56) are almost as effective as those of *A. mellifera* queens in inducing comb building in cages (Velthuis, 1976).

Queen pheromone and hoarding behaviour

Queen pheromones probably also stimulate worker bees to hoard food in the comb. Workers caged in groups of 50 collected more sugar syrup and stored more of it in small sections of comb when a queen was present than when they were queenless. The odour alone of the queen was ineffective (Free and Williams, 1972). The presence of a queen also stimulated caged pollen-gatherers to deposit their pollen loads in cells and to prepare empty cells to receive brood or food stores (Free and Williams, 1974).

Queen pheromone and foraging

Even when ample food stores are available in a colony bees are still stimulated to forage.

A queen's pheromones stimulate foraging in general and pollen-gathering in particular (Free, 1967b; Jaycox, 1970a,b; Free *et al.*, 1984a, 1985a). Within 24 h of the removal of a colony's queen, pollen collection is reduced and many of its pollen-gatherers change to collecting nectar only. A queen's absence also sometimes discourages nectar-gatherers from foraging, so although many pollen-gatherers may change to collecting nectar only, the amount of nectar-gathering does not always increase and may sometimes decrease, the overall effect being influenced by the original proportion of pollen-gatherers.

These changes in behaviour occur whether or not the colonies concerned contain brood; but the brood pheromones also have an important influence (see page 74).

It has sometimes been reported (e.g. Kashkovskii, 1957) that dequeening colonies increases honey production. This could be partly because the

Honeybee (*Apis mellifera*) foraging for pollen and nectar

queen's absence, and consequently less brood, results in a higher proportion of nectar- to pollen-gatherers, but probably mostly because a reduction in the amount of brood results in less nectar being expended on brood rearing.

Many beekeepers arrange to have two queens in their colonies for the greater part of the summer. This results in a greater honey yield than in single-queen colonies (e.g. Farrar, 1958; Moeller and Harp, 1965). The superior foraging performance of honeybee colonies that have been given two queens provides further circumstantial evidence of the stimulating effect of queen pheromones. In the past (Farrar, 1937) it has been concluded that honey production per bee in colonies with a single queen increases with colony population, so that strong colonies not only produce more honey, but more efficiently than smaller ones. Although recent experiments have cast doubt on this (Free and Preece, 1969; Barker and Jay, 1974) and have demonstrated that honey production increases in direct proportion to the colony population, the situation may be different in two-queen colonies. Two-queen colonies may produce more honey than single-queen colonies merely because they have more bees, but it is possible that the amount of queen pheromone per bee in two-queen colonies may also be greater and so the bees have more stimulus to forage and store pollen. An increased amount of queen pheromone per bee in two-queen colonies could also explain their diminished tendency to produce queen cells, supersede their queens and swarm (see Chaudhry and Johansen, 1971; Moeller, 1976).

A forager searching for a cell in which to deposit its pollen loads (*Apis mellifera*)

Experiments are needed to compare the activities of colonies of the same size, with one and two queens. Perhaps caging extra queens in a colony could increase the amount of queen pheromone available and so increase foraging. Tests should be made to find whether virgin queens of different ages and physiological conditions can be used effectively for this purpose.

It should be emphasized that although experiments have shown that a queen's absence diminishes foraging, it has yet to be demonstrated that an increase in queen pheromone in a colony with a mated laying queen would increase foraging. However, if the queen's pheromone that stimulates foraging could be synthesized and administered to colonies, pollen collection might be increased and the pollinating value enhanced, even of colonies already containing a queen.

Unfortunately the source and identity of the pheromone concerned is unknown. Jaycox (1970a) found that providing small confined colonies of 600 workers with 100 μg 9-ODA stimulated foraging for sugar syrup and partially compensated for the loss of their queens. But similar treatment for free-flying colonies of about 10 000 bees failed to influence the total number of foragers or those collecting pollen. Similar experiments by Youngs and Burgett (1982) and at Rothamsted Experimental Station, in which 9-ODA was administered to colonies, failed to elicit a response. This lack of success may partially reflect failure to attract bees to the synthetic pheromone, or to provide it in a form that the bees can collect and distribute.

Perhaps mated queens differ in their ability to stimulate foraging, reflecting differences in the amount of pheromone they produce; if this is found to be so it is an important factor to consider in breeding programmes. Young mated queens produce more 9-ODA than old ones (page 40). In New Zealand, Forster (1974) found that colonies with queens less than one year old produced an average of 30 lb of honey per colony more than those with queens of one or two years old.

Virgin queens stimulate foraging to a similar extent to mated queens, but are less effective at stimulating foraging for pollen (Free *et al.*, 1985a). Nevertheless, if they could be maintained in a colony possessing a mated and laying queen, foraging should increase.

Packages of bees in cheap containers, which can be destroyed when the pollinating task is complete, could be usefully employed in some circumstances, especially in locations where exposure to pesticide is severe. Virgin queens, which are cheaper to produce than mated queens should be effective in heading these units especially after being anaesthetized with carbon dioxide to induce them to lay eggs. However, the bees might build relatively more drone comb than when a mated queen is present; and perhaps as a result collect even less pollen and so pollinate too few flowers.

It would be more convenient if synthesized queen and brood pheromones could be used as substitutes for queen and brood in stimulating foraging and comb building. Probably disposable pollination units will not become a reality until these sythesized pheromones are available.

Pheromone produced from immature queens also stimulates foraging (Free *et al.*, 1984a). Queen cells, either open or closed, that were introduced to queenless colonies increased both nectar and pollen foraging, and were more effective than a mature queen in stimulating pollen collection. The latter effect may reflect the greater quantity of pheromone produced by the occupants of the five queen cells introduced to each colony than by a single mature queen. Even so, if the immature queen pheromone concerned could be isolated and synthesized it might be especially valuable in encouraging pollination of crops on which pollen-gatherers are the most successful pollinators.

Chapter seven

BROOD PHEROMONE: RECOGNITION AND STIMULATORY EFFECT

Brood recognition

In a honeybee colony, the ability of worker bees to recognize brood, its stage of development, sex and caste, is a prerequisite of brood care and progressive larval feeding. Cells containing brood are inspected at frequent intervals by house bees; presumably the stimuli they encounter release the appropriate behaviour patterns in physiologically and psychologically suitable bees.

Nurse bees inspecting larvae in their cells (*Apis mellifera*)

Cross-section of capped worker cells showing two pupae and polyethylene block

Free and Winder (1983) investigated the role of odour and other factors in the recognition of worker and drone larvae and pupae by determining the effects of various treatments, either in causing rejection of natural brood or in increasing the acceptance of 'artificial brood' consisting of small porous polyethylene blocks. The evidence obtained strongly suggested that a relatively involatile pheromone present on the body surface of brood is the primary brood recognition signal and worker bees need to make direct contact with the brood to perceive it.

Partially or completely masking the body surface of larvae with wax increased rejection; complete masking had the greatest effect, most larvae being removed after one or two hours. Washing larvae in solvent and so removing the recognition pheromone also increased rejection. Addition of the surface extract of honeybee larvae and pupae to polyethylene blocks introduced to empty worker cells diminished rejection (Fig. 7.1); it is hoped that a development of this technique will form the basis of a bioassay which will be useful in determining the chemical identity of the pheromone components concerned.

In a honeybee colony different diets are given to female larvae of different ages and castes, so workers providing the food must receive the appropriate signals, which are probably in addition to a more general brood-recognition pheromone. The amount of pheromone produced is likely to differ with a larva's size; this may help determine the amount of food it is given.

Immediately after larvae have been temporarily deprived of food they elicit more inspection and feeding visits than usual (Free *et al.*, 1987c). It is likely that the visiting worker bees are responding to a pheromone signal produced by the larvae. Perhaps other pheromones from the larvae signal the cessation of feeding and the initiation of building cell caps.

Drones are reared in larger cells than workers (page 79) and the differences in size (Koeniger, 1970) and odour (Free, 1967a, page 86) between the two types of cell probably enable the queen and workers to distinguish between them.

Nevertheless such differences are not exclusive; drone and worker larvae are readily accepted and reared when transferred to cells of the other caste (Free and Winder, 1983), drone-producing eggs are often laid in worker cells by queens without sperm and occasionally by laying workers (Free and Williams, 1974). It would be interesting to study the effects of introducing drone larvae or extracts of drone larvae to queen cells.

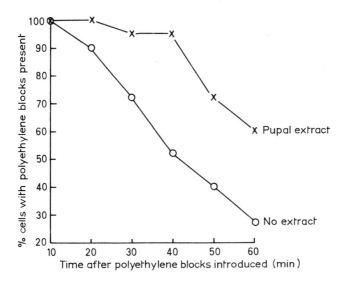

Figure 7.1 Effect of applying pupal extract to polyethylene blocks on their retention time in cells (after Free and Winder, 1983).

At certain times of the year, especially when insufficient nectar or pollen is available in the field, a proportion of eggs and very young larvae is often eaten by the workers so brood recognition pheromones can elicit different responses according to the circumstances. Always, proportionally more drone than worker brood is destroyed in this way (Free and Williams, 1975; Woyke, 1977). Any diploid drone larvae produced in an *A. mellifera* colony are usually eaten by workers within six hours of hatching. This behaviour appears to be released by a 'cannibalism' pheromone secreted by diploid drone larvae and so avoids the colony rearing unproductive diploid drones; normal larvae treated with extracts of diploid drone larvae are also eaten (Woyke, 1977). Secretion of the cannibalism pheromone of *A. mellifera* is restricted to the first one or two days after hatching. In contrast, diploid drone larvae of *A. cerana* are not eaten until two days old and some not until four days old, indicating that they secrete little or no cannibalism pheromone during the first day but thereafter secrete it for much longer (Woyke, 1980).

Probably the characteristics as well as the amount of the brood-recognition pheromone differ with the age of the brood. There is strong evidence that workers are able to differentiate between pupae of different ages. When caps are artificially removed from worker pupal cells the tendency to replace them increases with the age of the occupants (Fig. 7.2); within three hours of removing the caps of cells containing brood 18 days old or more, nearly all have been replaced (Newton and Michl, 1974; Free and Winder, 1983). When only a proportion of brood can be reared to maturity a primary concern for conservation of older brood is obviously in the best interests of the economy of the colony. Whereas the walls and bases of cells containing pupae

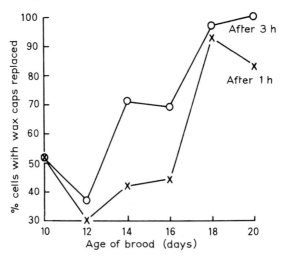

Figure 7.2 Relationship between the age of pupae and the tendency of bees to replace cell caps that had been removed experimentally (after Free and Winder, 1983).

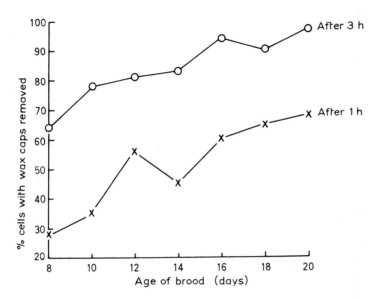

Figure 7.3 Relationship between the age of pupae experimentally removed from cells and the tendency of bees to remove the cell caps (after Free and Winder, 1983).

are covered with thin overlapping sheets of silk, the interior of the cell cap is covered with a thick mesh of threads (Jay, 1964). Presumably, the thick mesh allows sufficient odour permeability for workers to distinguish the presence, age and condition of the pupae inside.

When pupae were removed through openings made at the base of each cell, leaving the caps intact, the tendency of bees to uncap them increased with the age of the occupant removed (Free and Winder, 1983) i.e. as their former occupants were approaching the age at which the caps would have been removed naturally (Fig. 7.3). Perhaps therefore the wax caps of pupae cells incorporate features or retain odours that indicate the age of the occupants. It would be interesting to determine whether replacing older pupae with young ones for a day or so before leaving capped cells empty influenced the speed in which the cells are opened.

Koeniger (1978, 1984) found the order of preference for incubating different sorts of brood was: (1) closed queen cells, (2) capped drone brood cells, and (3) capped worker brood cells. He showed that both mechanical and chemical stimulation are involved in the incubation of queen cells. When he removed the pupae from queen cells the bees ceased to incubate them. However, when the pupae were immediately replaced with stones of similar weight incubation continued for about five hours but further incubation depended on the addition of pupal extract, presumably because the pupal pheromone present in the walls of the queen cell (page 14) had volatilized.

In a modified test (Koeniger and Veith, 1983) clustering bees were allowed to choose between two semi-artificial queen cells, consisting only of the natural silk cocoons. Treatment of one of the cocoons with the ethyl extract of 5–12 drone pupae induced bees to cluster on it in preference to the control cocoon. The main component concerned was identified as glyceryl-1,2-dioleate-3-palmitate and 7 μg of the synthesized component released clustering. This component occurs in worker, drone and queen pupae, (2–5, 10, and 30 μg in each respectively) but not in adult workers or drones (Koeniger and Veith, 1984). It also appears to be contained in olive oil; test cocoons impregnated with 0.4 μl olive oil released clustering (Koeniger and Veith, 1983).

Beekeeping applications: brood recognition

The sensitive and complex relationship between the immature and adult members of a honeybee colony results in the rapid perception and rejection of damaged brood. Even gently stroking larvae with a small brush increases rejection (Free and Winder, 1983). Recognition and rejection of diseased brood facilitates colony survival, and some colonies exhibit this trait more than others (Rothenbuhler, 1964; Milne, 1983a). A study of the factors releasing the eviction of diseased brood, possibly using the laboratory tests developed by Milne (1983b) might well give results of economic value.

Workers must be able to distinguish between worker and drone larvae (page 87) other than by the types of cell they occupy – this is most likely to be by different pheromone signals. The mite *Varroa jacobsoni* is an important parasite of honeybees in many parts of the world and is especially attracted to lay its eggs in cells containing drone brood (Ritter, 1981), probably recognizing the drone brood by the distinctive pheromone it produces. Development of a synthetic drone pheromone lure could well be useful in control measures.

Stimulatory effect of brood

The amount of pollen a colony collects increases with the amount of worker brood it is rearing (Todd and Reed, 1970; Barker, 1971; Al-Tikrity *et al.*, 1972). This may be partly because the amount of pollen collected directly influences brood rearing, and may be a limiting factor for colonies with little stored pollen (Fig. 7.4). However, the presence of brood in turn stimulates foraging, especially pollen collection (Fig. 7.5); the amount of foraging and pollen collected by colonies can be increased or decreased by adding or removing brood; these changes reflecting change in the foraging behaviour of the individual bees (Free, 1967b).

In one experiment removal of brood from colonies diminished pollen collection by 28% and removal of the queen two days later diminished pollen

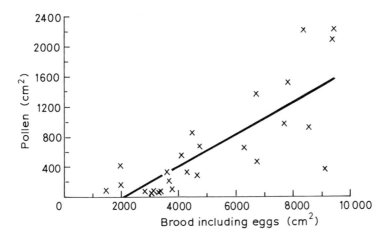

Figure 7.4 Relationship between the amount of stored pollen and the amount of brood in colonies (after Free and Williams, 1975).

collection by a further 27% compared to a control group of colonies that retained both brood and queen. At the beginning of another experiment three groups of colonies each had a caged queen but no brood. Brood was given to one group and the queens removed from the other. During the next three days the colonies given brood increased pollen collection by 223% and those whose queens had been removed reduced pollen collection by 62% (Free and Williams, 1971).

When foragers were excluded by wire screens from the brood and were only able to smell it, they collected less pollen than when they had access to it but more than when neither brood nor its odour were present (Fig. 7.6) (Free, 1967b). Therefore, although a volatile brood pheromone is partially effective in stimulating foraging, a contact brood pheromone seems of greater importance; this may be the same as the brood recognition pheromone described above. When foragers were able to make antennal contact, and possibly to transfer food, through a wire screen with bees caged on a brood comb, they collected more pollen than when they could merely smell the brood, but less than when they could touch it themselves. Presumably the bees caged with the brood transferred brood pheromone while they made antennal contact or gave food to those outside. Istomina-Tsvetkova (1958) found a direct relationship between the number of times a worker bee visited larvae and the number of times transfer of food occurred between it and other workers. Although this relationship presumably reflects the amount of food required for larval feeding, antennal contact during the food transfer probably helps to disseminate brood pheromones.

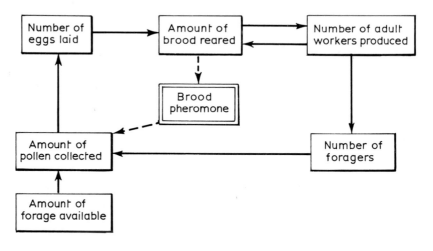

Figure 7.5 Influence of brood pheromone on pollen collection.

Honeybee eggs, larvae and pupae stimulate pollen collection, but larvae are most effective (Free, 1967b). When larvae are removed from bumblebee colonies (*B. pratorum*, *B. lucorum* and *B. pascuorum*) the amount of pollen collected is also greatly diminished (Free, 1955a). Jaycox (1970b) found that honeybee larvae may be relatively more important than the queen in stimulating pollen collection in poor foraging conditions. The results of attempts (Jaycox, 1970b) to stimulate foraging by providing colonies with a larval extract are difficult to interpret and in general were unsuccessful; the extract was less effective than larvae when dispensed to colonies in poor foraging conditions, but in good foraging conditions it appeared to stimulate pollen collection by free-flying colonies with queens.

It is not known whether the brood pheromones that stimulate foraging are the same as the brood recognition pheromones (Free and Winder, 1983) and the brood pheromone that inhibits ovary development of worker larvae (page 51), or whether worker and drone brood are equally effective.

Furthermore, it is not known why individual foragers differ in their responses to changes in foraging stimuli within their colony, so that some collect pollen only, some deliberately collect both pollen and nectar, some deliberately collect nectar and collect pollen incidentally, and some collect nectar only and may discard any pollen they collect incidentally. It does not seem to be related to the age of the bee or to the length of time it has been foraging.

Probably a bee's response to a particular stimulus depends partly on its physiological and behavioural past experience. A bee's response would also depend on the extent to which it is exposed to the stimulus concerned.

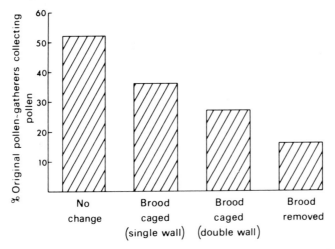

Figure 7.6 Effect of removing brood from colonies, enclosing it in a double-walled cage (so that bees on either side of the walls could not make contact), enclosing it in a single-walled cage (so that bees could make contact through the walls) and leaving brood uncaged on the proportion of foragers that continued to collect pollen (summarized from Free, 1967b).

Foragers that frequent the brood area of colonies and are exposed to brood pheromone are more likely to collect pollen than those that do not.

Beekeeping applications: stimulating pollen collection

When a hive entrance opens directly on to the brood area of a colony, a greater proportion of the bees using it collect pollen than when it opens onto an area of the hive with storage combs only (Free and Williams, 1976). Using two entrances, the lower adjacent to brood combs and the upper adjacent to storage combs, should therefore discourage the undesirable deposition of pollen loads in storage comb from which honey is to be extracted.

Not only the proportion of bees collecting pollen but also the actual amount of pollen collected can readily be influenced merely by altering the accessibility of the brood area to foragers (Free, 1979) (Fig. 7.7). On most crops, honeybees collecting pollen are more efficient pollinators than those collecting nectar only (Free, 1970a; McGregor, 1976), so to stimulate maximum pollen collection and pollination it is important that one or both entrances to a hive lead directly to nearby brood.

To pollinate crops in spring it is sometimes necessary to use small colonies whose brood area does not extend to near the hive entrance. In such circumstances it is possible to use simple devices to channel foragers from the hive entrance to the brood combs (Free and Williams, 1976). Further

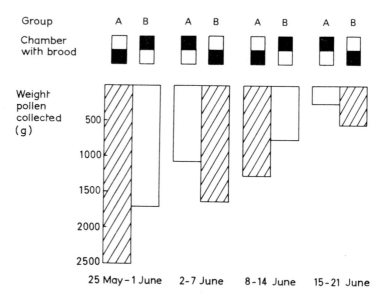

Figure 7.7 Influence of location of brood in hive on the amount of pollen collected by foragers. When the brood was in the lower chamber (group A on 25 May–1 June, 8–14 June; group B on 2–7 June, 15–21 June) the amount of pollen collected (cross-hatched) was always greater than when it was in the upper chamber. (After Free, 1979.)

advances will probably need to await the identification, synthesis and testing of the brood pheromones responsible for stimulating pollen collection. It is hoped that the pupal pheromonal component identified by Koeniger and Veith (1983) as glyceryl-1,2-dioleate-3-palmitate will prove effective.

Chapter eight

COMB PHEROMONE: RECOGNITION AND STIMULATORY EFFECT

Introduction

The wax secreted by bees and used by them to build comb in which food is stored and brood is reared could perhaps by itself be regarded as a pheromone. Furthermore, it is likely that pheromones incorporated into comb as it is built, or subsequently added to the cells may stimulate or regulate various activities of the colony. This is a subject that has been little explored.

The cells in the comb of a honeybee colony are hexagonal in cross-section and of two basic types. The larger cells (6.2–7.0 mm diameter) are used for the rearing of drone brood, and the smaller cells (5.3–6.3 mm diameter) for the rearing of worker brood (see Jay, 1963). Workers are produced from fertilized eggs and drones from unfertilized ones. The same cell may be used at different times for food storage and brood rearing. The brood often occupies a semicircular area of a comb, the top two corners of the comb and bands of cells at the side edges being used for storing nectar and pollen. Queen cells are vertical, circular in internal cross-section, and taper slightly from the base toward the open end.

A queen cell is increased in size as the larva inside it grows and is destroyed when the queen has emerged. In contrast, soon after an adult bee emerges from the drone or worker cell in which it was reared, the cell is cleaned by slightly older bees until its walls have a varnish-like coating before being re-used. The origin and purpose of any such material used to coat the cell walls is unknown and although the mandibular glands have been suggested as a source (Dreher, 1936) this seems unlikely (Callow *et al.*, 1959). Furthermore,

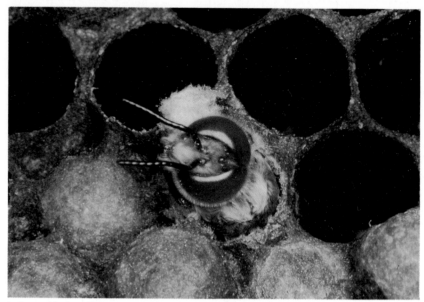

A drone (*Apis mellifera*) emerging from its cell

so far there is no evidence that bees prepare cells for particular purposes such as egg laying or pollen storage. Indeed a cell's function seems to be determined to a great extent by its position in relation to the natural configuration of brood and food stores in the hive (Free and Williams, 1974).

However, once cells are destined for brood rearing they are marked with pheromone. Cells from which eggs have recently been removed are preferred for food storage to cells that have remained empty; hence the bees obviously discriminate between these cells, and can only do so by differences in their odours. Possibly the eggs themselves deposit a pheromone that is adsorbed on the base of the cell, or more likely the queen or worker bees mark the cell for brood rearing once the eggs have been laid.

The immediate past history of cells greatly influence the retention of larvae transferred to them (Free and Winder, 1983). Survival is greatest in cells from which eggs or larvae have been removed and is least in new cells in which brood has never been reared.

Hoarding in worker comb

In the last few years there have been numerous studies of hoarding behaviour in which groups of 30–100 worker bees have been kept in small cages, each provided with a piece of empty comb and a supply of syrup and water. One factor influencing hoarding is the comb's previous use. Less syrup is stored in new comb than in old comb in which several generations of bees have been reared, or in old comb that has been used for storing pollen (Free and Williams, 1972). However, the tendency to prefer old comb may vary with different colonies (Rinderer and Baxter, 1980).

In another technique small sections of old and new comb were arranged in a chequerboard fashion within a normal comb frame, inserted into a hive and the use bees made of them compared (Free and Williams, 1974). Bees of the recipient colony preferred to store nectar and pollen, and especially the latter, in comb sections that had previously been used for storing food or rearing brood, rather than in the new comb.

When new, cells consist entirely of wax secreted by worker bees, but after brood has been reared in them they also contain cocoons and larval faeces; cocoons are light brown, so that after repeated use for brood rearing, combs become dark brown or black. Although attempts to show that substances associated with old cocoons release hoarding have not been successful (Free and Williams, 1974), this remains a possibility. Old comb is likely to have adsorbed a greater quantity of worker and queen trail pheromone (pages 108 and 16).

Empty comb, whether or not previously used for brood rearing, stimulates hoarding. Bees were stimulated to hoard more syrup in cages with three small pieces of comb than in cages with one piece only (Rinderer and Baxter, 1980) and used almost twice as many cells, even though the experiment was terminated while ample storage space still remained in each cage. In a later experiment the amount of syrup hoarded by groups of 30 caged bees increased as the area of comb provided increased from 47 to 281 cm^2 (Rinderer, 1982). Furthermore, the bees' hoarding response depended on the level of stimulus they had previously received, and was greater following a period of relative deprivation (Rinderer and Baxter, 1979).

The laboratory experiments on the stimulatory effects of comb have been complemented by field tests in which all the colonies used were provided with ample storage space but some had more than twice as much empty comb available as others (Rinderer and Baxter, 1978). During conditions of abundant nectar availability the colonies with most empty comb produced 30% more honey. Further studies (Rinderer and Hagstad, 1984) demonstrated that an increased amount of empty comb resulted in an increase in the proportion of foragers collecting nectar at the expense of pollen collection. This surprising result has important practical implications (Fig. 8.1). However, there is a possibility that giving colonies comb in excess of storage space likely to be needed, could result in partially filled combs, or in only some of the combs in honey storage chambers being used.

The exact nature of the stimulus from empty comb is unknown. In hives bees only occasionally visit comb that is not currently being used for storage, so it is unlikely that a contact response only is involved. Air that has been previously passed over empty comb stimulates hoarding (Rinderer, 1981), so probably a volatile pheromone is involved, the response increasing with the amount of pheromone. Perhaps the pheromone concerned is deposited when cells are prepared for use, and is obscured when they contain brood or stores. Air passed over comb containing honey does not increase hoarding. Providing honey producing colonies with synthetic pheromone instead of abundant empty comb would obviously be advantagous.

Exposing empty comb to isopentyl acetate, an alarm pheromone component from the sting chamber (page 139), diminished hoarding (Rinderer, 1982). This result is difficult to explain because bees soon become adapted to alarm pheromones (page 151) so it is unlikely to keep them in a state of permanent alertness. Perhaps it masked the pheromone present on empty comb prepared for food storage. In contrast, exposing empty comb to 2-heptanone, an alarm pheromone component from the mandibular glands (page 144), increased hoarding (Rinderer, 1982; Rinderer et al., 1982). Perhaps 2-heptanone may normally be used to mark cells for some purpose or other.

It is doubtful whether old comb has a special pheromone incorporated in it that stimulates foraging (Jaycox and Guynn, 1974), but experiments are needed to compare the nectar and pollen gathering of colonies based entirely on either new or old comb.

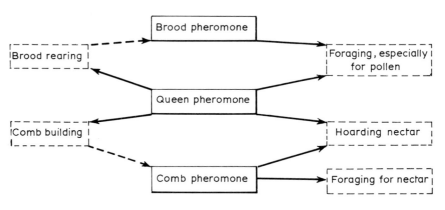

Figure 8.1 Stimulatory effect of brood, queen and comb pheromone.

Other stimulatory effects

Comb and the pheromones it contains may have stimulatory effects other than on hoarding and foraging. Jay (1975) showed that the presence of worker comb stimulated the development of ovaries of queenless worker bees kept in groups of 150 in small cages. Subsequent attempts to demonstrate this in small combless colonies were frustrated because the bees consistently built some comb themselves. Colony defence is also encouraged by the presence of empty comb within the nest (page 145).

Distinguishing between types of cell

Queen cups, the shallow dish-shaped precursors of queen cells, are available in the nest for most of the foraging season, but queens only lay eggs in them for some of this time, presumably after the workers have prepared them in some way which may include scent marking. However, no studies have been done on this.

Queens are able to differentiate between worker and drone cells, and lay in them fertilized and unfertilized eggs respectively. When a chequerboard arrangement of drone and worker comb is presented in the brood nest, the relative proportion of each used for brood rearing differs greatly at different times in the season, reflecting the amount of drone brood being reared under natural conditions (Free and Williams, 1974, 1975). Workers that have developed their ovaries and are laying eggs (unfertilized) prefer to do so in drone cells (Free and Williams, 1974).

Workers differentiate between drone and worker cells for purposes other than egg laying. When a small group of bees was caged with a piece of worker comb and a piece of drone comb they stored more syrup in the worker comb, although that stored in the drone comb was more concentrated (Free and Williams, 1972). When a chequerboard arrangement of drone and worker sections was presented in the hive, or empty drone and worker combs arranged alternately, the bees stored honey in worker cells in preference to drone cells, and stored hardly any pollen in drone cells (Free and Williams, 1974, 1975).

Queens are able to differentiate at least in part between worker and drone cells by measuring the width of the cells with their forelegs (Koeniger, 1970). Perhaps workers use a similar method to help distinguish between the two kinds of cells, both when laying eggs and storing food. However, it is likely that drone comb has a different odour from worker comb (page 86). Probably the pheromone concerned is the main means if not the sole one by which workers distinguish between drone and worker cells. Indeed, if it is true that worker and drone comb have different scents, this would greatly facilitate the work of the building bees. It would be interesting to know whether bees change freely between building the two types of cell, or whether they specialize in one type of cell only.

Chapter nine

REGULATION OF DRONE POPULATION

Introduction

Most honeybee colonies produce drones during spring and summer, and at the end of the summer drones are evicted from the hive by workers. The factors responsible for the initiation and cessation of drone production and the toleration or rejection of adult drones are far from clear but are probably under pheromonal control.

Drone cell production

Drones are reared in larger cells than workers (page 79) so drone rearing is usually only possible in a colony if it has drone cells. Under natural conditions colonies may build many drone cells, and in well-established wild colonies ample drone comb is usually present.

Bees of a swarm build only worker cells at first, and only later tend to build drone cells (Free, 1967a; Taber and Owens, 1970) so that at a certain stage in the growth of natural colonies drone cells tend to be located towards the edges and bases of the few large combs. As the colony grows the drone comb may be enveloped by worker comb built outside it.

The tendency to build drone comb at a later stage of comb building appears to be related to the size of colony population. When colonies were given empty frames in which to build comb the smaller colonies (up to 6000 workers) built no drone comb, or only a small proportion of drone comb whereas in larger colonies (10 000 or more workers) 80% or more of the cells built were usually drone (Free and Williams, 1975).

A drone (centre) surrounded by workers (*Apis florea*)

During winter, as the bees form a compact cluster, they withdraw from the cells at the outside of the combs; in the following spring, as they become more populous, they spread out to occupy them again, and if necessary build more comb. In England the proportion of drone cells built is greatest in the three summer months of May, June and July although colonies continue to build drone comb long after they cease to rear drones (Free and Williams, 1975).

The reluctance of colonies to build drone comb in early spring when their worker populations are relatively small, and the reluctance of small colonies to build drone comb at any time, suggest that pheromones from the queen help inhibit drone cell production, but when a colony becomes larger and less queen pheromone is available per bee, inhibition is no longer complete. Old and failing queens produce less pheromone (pages 40 and 44); it has long been supposed that their colonies build much drone comb but this needs to be verified. Conversely, it is reported that drone comb is seldom built in colonies headed by queens reared in the current year (Wedmore, 1932) – this too needs verification.

Although diminution in the amount of queen pheromone per bee may initiate drone cell construction its suppression does not depend solely on an increase in queen pheromone availability, because when the proportion of drone comb built diminishes toward the end of the summer the worker populations of the colonies are still large.

The source of any queen pheromone involved in inhibiting drone cell construction is unknown; perhaps it is other than the mandibular glands as

Zmarlecki and Morse (1964) found that drone populations remained normal in two colonies that were headed for eleven months by queens whose mandibular glands had been extirpated.

Female brood also influences the amount of drone comb built. For the first few days after combs had been removed from colonies only worker cells were built. The queens were then removed which greatly diminished the rate of cell building but still all those built were worker cells. Each colony was then given a comb of worker larvae and during the next few days all built much drone comb (Free, 1967a). However, because on each comb that had been introduced a few of the female larvae were being reared as queens, further experiments are needed to compare the effects of queen and worker larvae on drone cell production. The influence of a virgin queen also needs to be investigated.

Darchen (1968) confirmed the inhibiting effect of a queen on drone cell building. The half of a colony that retained its queen built only worker comb while the queenless half built drone comb. He obtained evidence that the presence of worker pupae encourages drone cell construction. Queenless colonies fed sucrose syrup built drone cells only when they had combs containing worker pupae, but when they had no worker brood they changed to building worker comb.

The amount of drone comb built also depends on the amount already present. It is possible to reduce the number of drone cells built in a colony by adding drone comb, or to increase drone cell construction by removing all existing drone comb (Free, 1967a). It seems likely that an identifying pheromone is incorporated into drone or worker comb as it is built (page 71). Experiments are needed to determine whether the odour alone emanating from drone comb has an inhibitory effect.

Drone brood production

The amount of drone brood and the number of adult drones in a colony is also correlated positively with the size of its worker population (Mathis, 1947). Colonies with little brood have only a small proportion of drone brood or none; colonies with much brood tend to have the greatest proportion of drone brood (Free and Williams, 1975).

Although in England a large percentage of eggs are laid in drone cells before the end of April, few are reared, and the proportion of drone brood in a colony is at its maximum in May and June (Free and Williams, 1975). Queen cells are produced most abundantly during late June and early July, but within a colony there is no clear evidence of correlation between the dates of drone brood and queen cell production (Simpson, 1957; Allen, 1965b). Furthermore, there is no evidence that drone rearing releases queen production.

A drone being evicted from its hive by workers (*Apis mellifera*)

When drone production is limited by shortage of drone comb much of the drone comb built is used immediately for rearing brood. In these circumstances it is probable that the building of drone comb and the rearing of drone brood are governed by similar factors. When more than sufficient drone cells are available, the workers appear to help regulate the amount of drone brood reared by destroying drone eggs or young larvae, especially in small colonies and early in the season.

It seems that a pheromone produced by the drone brood itself is partly responsible for this regulatory mechanism. Removing drone brood from colonies encourages its production, and adding drone brood to colonies has the reverse effect (Free and Williams, 1975). The adult drone population is also influential. Groups of 0, 500, 1000 or 2000 adult drones were added to large colonies containing no drones but empty drone comb. The number of drones reared in these colonies was inversely proportional to the number of drones added (Rinderer *et al.*, 1985).

Eviction of adult drones

The eviction of drones by the workers is seldom a rapid process. Before drones are expelled they are often denied access to the honey stores and become too weak to avoid being dragged from the hive.

One of the most important factors initiating the seasonal rejection of drones seems to be the amount of forage being collected. If this is terminated drones are still evicted, whereas when it is prolonged beyond the normal seasonal span drone eviction is curtailed. An abrupt change in the supply of forage produces a more marked response than a gradual one; in dearth conditions drones may even be expelled in June. At any time of the year a colony can be caused to evict its drones by preventing it from foraging.

However, lack of forage only releases eviction when the colony has a queen. Even at times when little forage is available no drones are evicted from colonies without mated laying queens; in such colonies eviction begins soon after the virgin queens they have reared have mated and laid eggs. This is probably in association with the increased amount of pheromone the queens are producing (page 25) but may be released by mating or egg laying. Observations on drone eviction in colonies with virgins that are prevented from mating could help to clarify the issue. In autumn eviction can be greatly delayed by providing colonies with artificial forage and by removing their queens; these effects are additive (Table 9.1; Free and Williams, 1975). It would be interesting to know whether colonies that have evicted their own drones would be ready to accept others if foraging conditions improved or if they became queenless.

Table 9.1   Effect on drone eviction of feeding colonies and of removing their queen (after Free and Williams, 1975).

Group	Number of drones on 6 August	Number of drones on 2 September	% Evicted by 2 September
Queens present, colonies not fed	1192	0	100
Queens present, colonies fed	1480	150	90
Queens absent, colonies not fed	2273	1098	52
Queens absent, colonies fed	1330	825	38

Regulatory mechanism

The production of drone comb and brood and the toleration of adult drones in a colony seems to be regulated by a balance between stimulatory and inhibitory pheromones (Fig. 9.1). Probably a laying queen produces pheromone that inhibits the rearing of drone brood and the tolerance of drones. It seems likely that worker brood produces a pheromone that stimulates drone production and that the relative amounts of queen and brood pheromone determine the proportion of drone cells built, the amount of drone brood reared and the attitude of workers toward drones. Thus, when a queen is removed from a colony drone cells are built only if worker brood is present.

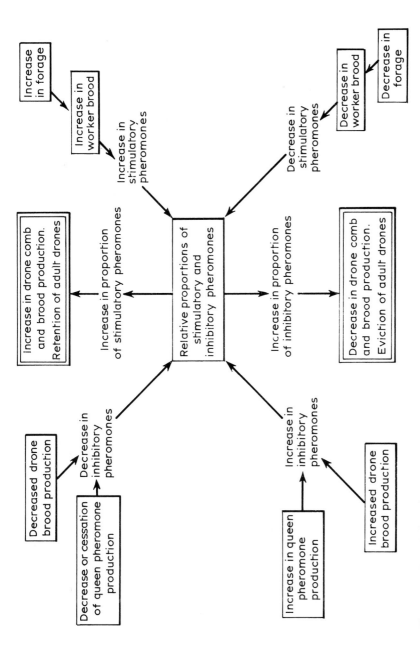

Figure 9.1 Regulation of drone production.

The reluctance of colonies to build drone comb in early spring when their worker populations are relatively small, and the reluctance of small colonies to build drone comb at any time, suggest that the amount of queen pheromone is sufficient to inhibit drone cell production in small colonies but as a colony becomes larger it is no longer sufficient. Early in the season the small amount of worker brood present in a colony produces insufficient pheromone to counteract the queen pheromone but as the amount of worker brood increases there is also an increase in the proportion of drone brood reared. Toward the end of the summer, as the amount of worker brood reared diminishes, the queen's inhibitory pheromone reasserts its influence. Drone cell production, drone rearing and the tolerance of drones by workers is encouraged in the presence of failing queens with a diminished output of pheromones. Furthermore, drone comb, drone brood and adult drones also appear to produce inhibitory pheromone; this would account for the feedback mechanism whereby when sufficient drone comb, drone brood or adult drones are present further production is inhibited.

A steady diminution in the amount of forage available would cause a corresponding reduction in the amount of worker brood reared and hence of the tolerance of adult drones. It is more difficult to explain the release of worker hostility to drones following an abrupt cessation of foraging. Perhaps foragers are basically hostile to drone odour but in the presence of ample forage this underlying hostility is repressed; an abrupt end to foraging leaves numerous idle foragers who are ready to defend their colony against intruders with hostile odours (page 102). Furthermore, foragers are less likely than nest

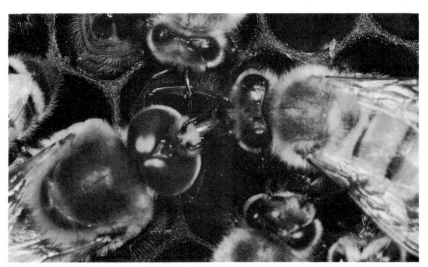

A drone (left) receiving food from a worker (right) (*Apis mellifera*)

bees to receive the relevant pheromone emanating from worker brood, which could help to explain why older drones may be attacked by unemployed foragers while younger drones of the same colony are being fed by the workers. This hypothesis needs to be tested experimentally by varying the relative amounts of the queen and worker brood pheromones present.

Beekeeping applications

Many beekeepers regard drone production as wasteful and attempt to avoid having any drone comb in the brood area of colonies. Because the amount of drone comb in a colony overrides the effect of the time of year in determining whether more is built, presumably through the amount of pheromone produced, using drone comb in the honey storage supers should prevent drone comb being built in the brood area. However, although the absence of pollen in the drone comb provided would be an additional bonus, bees do not favour drone comb for honey storage – so the use of drone comb in the honey storage supers of beehives is of questionable value. If the appropriate pheromone of drone comb could be isolated and synthesized it might be useful in inhibiting drone production in colonies that could also be kept free of drone comb.

Darchen et al., (1957) stated that no drones appeared in multi-queen colonies. If the role of queen pheromone in inhibiting drone cell production is verified it should eventually be possible to increase the amount available per worker by introducing additional queens (page 66) or synthetic queen pheromone.

Drones from Africanized colonies in South and Central America readily join and adopt colonies of European bees, whereas the reverse occurs much less often. As a result there is a deficit of drones in Africanized colonies and they are encouraged to produce more, but the influx of Africanized drones into European colonies inhibits drone production. The Africanized colonies therefore have an important reproductive advantage. Where appropriate, efforts could be made to exclude Africanized drones from European colonies and destroy them (Rinderer et al., 1985).

Chapter ten

MATING PHEROMONES

Introduction

During mating drones and queens occupy 'air space' at 10–40 m above ground, which is considerably higher than that usually used by foraging workers. This makes the study of mating behaviour difficult. However, as early as 1902, F.W.L. Sladen made some very astute observations:

'The hum of drones flying in the air can often be distinctly heard, especially in open country where the ground is high or rising. In such a spot, on three or four successive warm and sunny days in the early part of July 1902 I found it quite easy to attract swarms of drones to a virgin queen that was placed in a cage on a post about 4 ft above the surface of the ground, between 1 and 3 p.m. The drones were attracted also by the cage alone after the queen had been in it. Other drones were attracted to these drones, chasing one another, and sometimes a little swarm of drones would fall to the ground. When several virgin queens were exposed a short distance from one another the drones generally paid attention to one only. The attractiveness of a queen was increased by association with the drones, and also by raising her 20 or 30 ft by means of a kite. It is clear that the scent of the queen attracted the drones, and it seems to me likely that the ardent drones attracted the other drones by emitting a scent themselves' (Salmon, 1938).

Role of pheromones in mating

Mating behaviour received renewed interest when Gary (1962) developed a drone attraction bioassay in which tethered queens were suspended, at heights of 10–25 m, from helium filled balloons or from a horizontal nylon line stretched between two 14 m poles. Each queen had a nylon tether, approximately one metre long, glued to her thorax so she was able to fly in a small circle. When the queen was raised aloft drones were soon attracted to her.

Small pieces of filter paper on which virgin queens had been squashed also attracted drones. The source of this attraction was shown to be the queen's mandibular glands. Synthetic 9-ODA, the main component of the mandibular glands (page 21), was by itself effective. However, the individual fraction of the mandibular gland secretion that contained 9-ODA was less attractive than the whole complex.

Adaptations of Gary's technique have been used by several other workers: pheromones have been presented on cotton wool, filter paper or small porous polyethylene blocks together with lures which were either dead extracted queens or workers or small wooden 'models'.

Newly emerged queens have little 9-ODA but the amount increases until they are about ten days old (Butler and Paton, 1962; Pain et al., 1967; pages 21 and 23). Probably in association with this, virgin queens do not attract drones until they are five days old and elicit a maximum response when they are about eight days old (Butler, 1971). Production of 9-ODA in virgin queens is greatest from 11.00 to 17.00 h daily (Pain and Roger, 1978); during this time of day mating flights occur.

Special olfactory sensory cells on drone antennae respond to 9-ODA and to the mandibular gland secretion of queens (Beetsma and Schoonhoven, 1966; Boeckh et al., 1965; Ruttner and Kaissling, 1968). Response appears to be confined to the *trans* isomer of 9-ODA, the *cis* isomer having no influence (Doolittle et al., 1970; Adler et al., 1973). Drones are much more sensitive to the *trans* isomer than either workers or queens, responding to as little as 0.025 µg; moreover with increase in the amount of 9-ODA (up to about 50 µg) the antennal response of drones increases proportionally, whereas the other two castes maintain a similar level of response.

Some authors (Boch et al., 1975) have found laying queens are more attractive than virgin queens and associated this with the larger amounts of 9-ODA found in their mandibular glands (i.e. averages of 285 µg for laying queens and 150 µg for virgin queens), but other authors (Pain and Ruttner, 1963; Butler and Fairey, 1964; Butler, 1971) have found virgin and mated queens are equally attractive.

Pain and Ruttner (1963) confirmed Gary (1962) and reported that groups of drones in the vicinity of a lure treated with an extract of a queen's mandibular glands were larger and more stable than groups near lures with 9-ODA only, whereas Butler and Fairey (1964) and Renner and Vierling (1977) were unable to discover any difference between the attractiveness of a queen equivalent of 9-ODA and a complete extract of a virgin or mated queen. Addition of 100 μg 9-ODA to a queen whose mandibular glands had been removed restored her attractiveness to normal (Butler, 1971).

However, when comparing a given amount of 9-ODA with a queen's body or extracts of a queen's body the results are difficult to interpret; it must be borne in mind that there is little information (page 50) on the amount of 9-ODA normally present on the body surface, and on its rate of release from the mandibular glands and body surface. The release rate of 9-ODA from the body surface is probably reduced by the presence of a 'keeper' substance, so that although lures with synthetic 9-ODA alone may be initially more attractive to drones, lures with queen extracts containing equivalent amounts of 9-ODA remain attractive for longer.

Butler and Fairey (1964) discovered that 9-hydroxy-2-decenoic acid (9-HDA), another major component of the queens mandibular gland secretion (page 22), was somewhat attractive to drones, although less so than 9-ODA, and the attractiveness of 9-ODA was not increased by the addition of 9-HDA; Blum et al., (1971) and Boch et al., (1975) were unable to confirm the attractiveness of 9-HDA.

Hence the subject is controversial and any components in the mandibular gland secretion that increase the attractiveness of 9-ODA still await identification. At any one time the number of drones flying, the height of flight, their responses to lures and discriminating ability depend on many variables including wind speed, wind duration, temperature, terrain and local population density. No doubt these variables have contributed to the conflicting results. Simultaneous comparisons of the different treatments under investigation should always be made. Gary (1963) suggested that in bioassays to detect any weakly attractive substances it would be useful to have a bait, such as a tethered queen or whole queen extract, to lure drones into the vicinity.

Observations (Gary, 1963; Butler and Fairey, 1964) indicate that drones are attracted upwind to the odour of a queen (of which 9-ODA forms a major part) from a distance of up to about 60 m. On reaching a tethered queen or a queen model they hover slightly below and to leeward of her and examine her abdomen with their antennae and forelegs. They can probably see her from about one metre distance. The groups of drones hovering at a queen or lure often adopt a comet-like formation whose shape is continually changing in accordance with variations in chemical and visual stimuli. Drones immediately downwind of a lure containing excessive amounts of 9-ODA may themselves attract other drones, presumably because they have become contaminated with 9-ODA.

If a drone flies past the queen odour source it will turn back after about 10 m and search for the scent again. Within limits, an increase in wind speed can result in more drones finding the tethered queen. In one experiment (Gary, 1963) queens were suspended in a row parallel with the direction of the wind. Groups of drones flew windward, stopping momentarily around each queen; when they had passed the last queen in the row they flew back to the beginning.

The queen's sting chamber must be open for copulation to occur; probably drones establish whether this is so while palpating her abdomen. Gary (1963) attributed the lack of successful mating of tethered queens to their failure to open their sting chambers. He induced drones to mate with queens whose sting chambers had been forcibly fixed in an open state, with queens whose abdominal segments 7–10 had been removed leaving the body cavity exposed to represent an artificial sting chamber, and with queens provided with an artificial aluminium foil sting chamber. In later experiments (Gary and Marston, 1971) copulation was achieved using wooden models provided with queen odour and an orifice of suitable depth and diameter (Fig. 10.1). Butler (1967) confirmed that both 9-ODA and an open sting chamber were required to elucidate mating; neither on its own was sufficient. Whereas a queen with a closed sting chamber received as many drone visits as one with a sting chamber open, more drones attempted to mount the latter. It seemed unlikely that the sting chamber itself produced a mating pheromone; indeed, freshly killed worker bees with open sting chambers and treated with 50 μg 9-ODA were almost as effective as freshly killed virgin queens.

However, more recently it appears probable that when a drone is hovering to leeward of a queen the odour from her tergite glands also helps to induce him to mate. Renner and Baumann (1964) suggested that the volatile

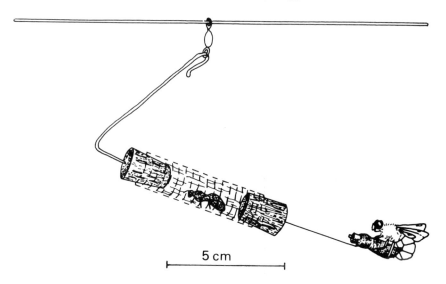

5 cm

Figure 10.1 Drone mating with wooden model after being attracted by pheromone applied to model and by pheromone from caged virgin queen (from Gary and Marston, 1971).

pheromone from the queen's tergite glands (pages 18 and 20) might be important in the mating process because the odour of it is very noticeable in virgin queens of one to two weeks old. This was demonstrated by Butler (1971) – the number of times drones seized dead 2-day old queens and attempted to mate with them was increased both by the addition of synthetic 9-ODA and by rubbing their abdomens with the abdominal tergites of 12-day old virgins; the presence of synthetic 9-ODA alone increased the attractiveness of the 2-day old virgins to drones but did not induce seizure. Further experiments (Renner and Vierling, 1977) suggest that whereas the queen's mandibular gland secretion is responsible for attracting drones to her vicinity, within 30 cm, pheromones from her tergite glands are predominantly attractive and release attempts to mate; when small paper tubes 3 mm in diameter were treated with tergite gland secretion drones attempted to mate with them. Because the tergite gland secretion exerts its influence in stimulating copulation, after the queen and drone have been brought together by the mandibular gland sex attractant, it may be regarded as an aphrodisiac.

Morse et al., (1962) removed the mandibular glands from six 3-day old virgin queens and subsequently two of them mated; possibly therefore the secretion from the tergite glands was sufficient to allow these matings to occur. This needs further investigation.

Drone congregation areas

Zmarlicki and Morse (1963) established that drones could be attracted to elevated tethered queens in some areas but not others, and so produced the first real evidence that there are special areas where drones congregate and to which virgin queens fly to mate. Further work, especially by Ruttner and Ruttner (1965, 1966, 1968), has helped to provide much information about drone congregation areas but there are still many unanswered questions.

A drone congregation area may be defined as an air space 15–25 m above ground, where drones habitually collect and fly around, independently of the presence of queens, and to which queens come to mate. The area itself seems to have an astonishing power to attract and retain drones. Ruttner (1966) observed that even drones which had located a tethered queen left her when she was moved out of the area and they flew back into it. This is not always so, as in South Africa A. mellifera scutellata drones would continue to follow an elevated lure for 2 km (Tribe, 1982).

Congregation areas occur especially in hilly or mountainous country. However, their characteristics have proved surprisingly difficult to define and more research is needed. They can be above flat clearings surrounded by trees, an open hollow or a slight summit. While the criteria for a drone assembly area are not properly understood, it appears that in hilly regions drones can usually reach them by making an uninterrupted straight flight from their hives in the direction of the lowest point on the horizon. Strang (1970) gained the impression that the areas themselves are marked by some form of vertical relief such as a hedge, trees or the side of a building.

The limits of drone congregation areas have been established by elevating a queen or pheromone lure from a balloon, or by towing a lure behind a radio controlled model aircraft (Gerig and Gerig, 1976), and observing the number of drones visiting it.

The size of drone congregation areas vary considerably, from 30–200 m across, in different locations, although neither the dimensions nor locations change greatly from year to year. Indeed, congregation areas persist for several consecutive years. One, which is still in use, was described as follows by Gilbert White at Selborne, England in 1768 although its purpose was not then appreciated:

'There is a natural occurence to be met with upon the highest part of our down in hot summer days, which always amuses me much, without giving me any satisfaction with respect to the cause of it; and that is a loud audible humming of bees in the air, though not one insect is to be seen. This sound is to be heard distinctly the whole common through from the Money-dells to Mr. White's avenue gate. Any person would suppose that a large swarm of bees was in motion, and playing about over his head. This noise was heard last week, on June 28th'.

There may be as many as 10000 drones in a single drone congregation and they come from many colonies from different directions and up to 5 or 6 km distance, although the average is nearer 1 km. When colonies are moved to a new location their drones quickly locate the congregation areas.

Tribe (1982) reported that during calm periods drones disappeared from congregation areas and reappeared after the wind had increased. In the absence of wind drone flight may cease because of difficulty in locating flying queens, but wind direction may also be important in enabling drones to locate a congregation area. Tribe (1982) supposed that the presence of wind turbulence at a site may encourage the circulation of any queen and drone pheromone released there, and, furthermore, that on leaving their hives drones fly mainly upwind until they encounter a turbulent area where they await a queen.

There is little information as to how queens find drone congregation areas. If a queen flies upwind on leaving her hive she will leave a much longer odour trail for drones to follow than if flying with the wind. Drones are most likely to find her if she hovers in one place, approximately stationary, or flies around in a small area, as at a drone assembly site. Drones themselves may produce an odour that attracts queens and could help guide new drones to a congregation area. Gerig (1972) reported that extracts from drone heads attract flying drones once they have arrived at a congregation area; if this is confirmed it is possible that a drone pheromone enhances or maintains comet formation.

Once the queen has arrived at a drone congregation site, the presence of an abundance of drones together with efficient pheromonal communication must ensure that she mates with several drones in quick succession. This must help reduce the time she is exposed to predator attack. Furthermore, because each congregation site is visited by drones from a large catchment area, the chances of inbreeding are reduced.

In flat, featureless countryside drone congregation areas may be absent. In such circumstance a meeting of queen and drones must be more difficult to achieve, and the queen probably needs to fly out more than once to accomplish the required number of matings, with attendant risks. This has yet to be demonstrated. Perhaps on flight paths to the assembly areas an intermediate situation occurs.

Other *Apis* species

Attempts have been made to attract drones to alkenoic acids closely related to 9-ODA but without any success (Blum *et al.*, 1971), indicating that the honeybee sex attractant has a high structural specificity which most effectively eliminates attempted matings with species other than honeybees. However, the mandibular glands of queens of all honeybee species contain 9-ODA (page 21); *A. mellifera* is naturally isolated geographically from the remainder, but *A. cerana*, *A. dorsata* and *A. florea* are often sympatric.

Butler *et al.* (1967) found no difference in attractiveness of ethanol extracts of *A. cerana*, *A. florea* and *A. mellifera* queens to *A. mellifera* drones. Extracts of *A. dorsata* were not available to them, but later Shearer *et al.* (1970) found that *A. mellifera* drones were also attracted to extracts of heads of *A. dorsata* queens, and indeed prefered them to lures containing 100 µg 9-ODA. In similar tests Sannasi *et al.* (1971) found that extracts of heads of *A. cerana*, *A. dorsata* and *A. florea* attracted 505, 425 and 510 *A. mellifera* drones respectively compared with 315 to a polyethylene lure dosed with 100 µg 9-ODA. Perhaps this was because the queens contained more 9-ODA than the lures (about 200+ µg for queens 6 weeks old or more, of *A. cerana*, *A. dorsata*, *A. mellifera* according to Shearer *et al.* (1970) – which is considerably more than the 130 µg and 110 µg found by Butler and Paton (1962) and Pain *et al.* (1967) respectively for *A. mellifera* queens), or perhaps because the mandibular glands of all species contained attractive material other than 9-ODA.

Drones of all three species of tropical honeybees are also attracted to *A. mellifera* queens and to synthetic 9-ODA (i.e. *A. dorsata* in the Philippines, Shearer *et al.* (1970); *A. cerana*, *A. dorsata* and *A. florea* in India,

Sannasi *et al.* (1971)). It is likely that there is inter–attraction between them, although experiments to show this do not seem to have been done. Because of disparity in size and other anatomical differences, interspecific mating is unlikely to occur, but interspecific attraction alone could delay and, possibly prevent, natural mating.

Perhaps in addition to 9-ODA each species produces a specific pheromone component; experiments are needed to examine this possibility. There is some indication that this might be so. Ruttner and Kaissling (1968) found that in field trials *A. mellifera* drones were attracted to both *A. mellifera* and *A. cerana* queens but on some days the *A. mellifera* queens were preferred. In the Philippines attempts to attract *A. cerana* drones to 9-ODA alone met with little success.

Differences in the flight times of the drones and queens of different species could help achieve ethological isolation in sympatric species. In Sri Lanka the daily periods of drone flight are: *A. florea*, 12.00–14.30 h; *A. cerana* 16.15–17.15 h and *A. dorsata* 18.00–18.45 h (Fig. 10.2; Koeniger and Wijayaguna-sekera, 1976). Therefore, assuming that the queens fly only during the same periods as the drones any attempts at interspecific mating would be avoided. These flight times for the different species are not adhered to elsewhere, but it would be interesting to determine whether separation of flight periods also occurs in other regions.

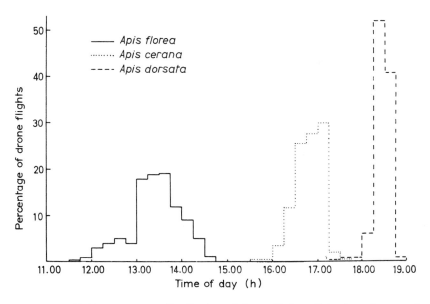

Figure 10.2 Daily percentage distribution of drone flights of the honeybee species *Apis florea, Apis cerana* and *Apis dorsata* in Sri Lanka (after Koeniger and Wijayaguna-sekera, 1976).

It is sometimes difficult to maintain colonies of *A. mellifera* when they are first introduced to an area where other *Apis* species occur. Presumably, in these circumstances, the flight times of *A. mellifera* queens and drones overlap those of one or more of the other *Apis* species present. If so, the *A. mellifera* queens could well be harassed by the numerous *A. cerana*, *A. dorsata* and *A. florea* drones and fail to mate with the relatively few *A. mellifera* drones, many of which would be diverted to queens of other species.

When *A. cerana* colonies were introduced from Pakistan to West Germany their virgin queens visited *A. mellifera* drone congregation areas; mating could only be achieved after separating the two species by 10 km (Ruttner et al., 1972; Ruttner and Maul, 1983).

Beekeeping applications

Synthetic mating pheromones can be used to survey the flight distribution patterns of drones of different genetic origins, so that attempts can be made to locate colonies containing virgin queens in areas where the most desirable matings are likely to be achieved. The drones can be captured in sweep nets after lowering elevated lures, or they can be trapped in nets towed behind radio controlled model aircraft equipped with lures (Gerig and Gerig, 1976).

It might prove possible to localize mating near selected drone stocks by establishing artificial drone congregation areas. Perhaps baiting a site with large amounts of synthetic 9-ODA or drone pheromone could achieve this. However, it has yet to be shown that queens would naturally fly into such areas, and if they do so whether males can find them amidst the confusion of abundant sex pheromone in the environment. Virgin queens could be tethered and suspended in an artificial drone congregation area – however they seldom open their sting chambers and mate when tethered (Gary and Marston, 1971). Attempts have been made to overcome these difficulties by forcing open the sting chambers of tethered queens with rings, wire springs, hooks or glue and by attaching the tether to the queen with a glue which the workers can remove when she is returned to her hive (Butler, 1967, 1971; Gary and Marston 1971; Koeniger *et al.*, 1979a).

Artificial mating stations could perhaps be of value in attempting to change the characteristics of the feral population. Possibly they could be used to attract virgin queens of aggressive Africanized stock in South and Central America to mate with a docile strain of drone.

Chapter eleven

NEST AND NESTMATE RECOGNITION

Introduction

Honeybees use the odours emanating from their colony to help them locate the entrance to their nest on return from the field, and use the odours emanating from individual bees to help them distinguish friend from foe.

Nest odour and nest recognition

Foraging honeybees that have found the entrance to their hive after being temporarily disorientated release Nasonov pheromone (page 114) at the hive entrance. Odours from inside the hive that release this behaviour include that of the comb, food stores, adult workers, drones, and queen (Ferguson and Free, 1981). Presumably all bees respond to these combined odours when returning home from field excursions, and together they may be regarded as the hive or nest odour.

However, bees seem to take little note of any odour specific to their own hive and visual orientation seems of much greater importance. Free (1958) allowed foragers to choose between their own hive and another which was identical to and equidistant from the original site. Presumably the only means by which they could distinguish their own hive from the other was by its distinctive odour, but they entered their own and the new hive in about equal numbers indicating that any difference in hive odours was of little significance. Butler *et al.* (1969) also showed that returning foragers respond to a general hive odour but found that they had little preference, if any, for trail odour left at the hive entrance by bees of their own rather than of a strange colony.

Although specific hive odours seem to be of little consequence to orientating bees, they are able to differentiate between the odour of their own and foreign hives; it has been demonstrated that odour pumped from a foreign hive is less effective than that from their own hive at inducing worker bees at the hive entrance to expose their Nasonov glands (Ferguson and Free, 1981).

Colony odour

Honeybees need to defend their food stores from robber honeybees of other colonies, especially at times of the year when little forage is available in the field. It is therefore of vital importance that guard bees should have a means of distinguishing between members of their own and other colonies. It has long been supposed that worker bees of the same colony share a common distinctive colony odour which is different from that of other colonies. Proof for this was provided by foraging experiments (von Frisch and Rösch, 1926; Kalmus and Ribbands, 1952) which showed that the foragers of a colony preferred to visit glass dishes containing sucrose syrup on which members of their own colony had foraged to dishes on which bees of another colony had foraged, and by experiments at the hive entrance (Butler and Free, 1952; Ribbands, 1954; Free, 1954) which showed that guard bees recognize intruders by their alien odour. These odours appear to be relatively non-volatile and only perceptible by contact chemoreception.

When workers recently killed by cold are introduced into the top of a hive they are evicted through the hive entrance sooner if they come from an alien

Worker honeybees (*Apis mellifera*) attacking a stranger at the entrance to their hive

colony than if from the recipient colony. Furthermore, bees are able to distinguish between the heads alone of bees of their own colony and heads of bees of alien colonies; they prefer to beg food from the former and may even attack the latter (Free, 1956).

Even colonies in the same location make different use of the local flora, and Kalmus and Ribbands (1952) argued that slight differences in the quality of forage being used in different colonies was sufficient to impart differences in colony odour. They supposed that because of the extensive transfer of food between the members of a colony, each one would have an almost identical diet, would ingest a similar mixture of the nectars collected, and that odorants associated with the mixture would be passed to the epicuticle and form the basis of the colony odour. Although at any one time all the workers of a colony would emit a similar odour, this would change with the colony's incoming food supplies. In support of their hypothesis they demonstrated that a change in colony odour could be induced by a rather drastic change in food supply. Dividing a colony into two equal parts and feeding one part with a mixture of black treacle and heather honey soon resulted in it acquiring a different odour from the unfed part. Conversely, Ribbands (1953) found that taking colonies (whose food stores he had removed) to an area where heather was the only source of forage, resulted in all the colonies acquiring the same colony odour and the bees being unable to distinguish between members of their own and other colonies.

An alternative suggestion (Butler and Free, 1952) was that distinctive colony odours arise by adsorption of odours present in the hive atmosphere onto the waxes of the body surface. This alone could have accounted for changes in colony odour when, in the above experiments, colonies were given food with a strong aroma. It would also explain why intruders that had managed to remain in alien colonies for only one or two hours appeared to have acquired the odour of the bees of the adopted colonies and were no longer treated as alien.

Experiments by Renner (1955, 1960) provided evidence for this suggestion. He showed that the odour alone of black treacle in a hive was enough to change the colony odour even when the bees could not feed on it. He also obtained Nasonov pheromone, free from other odours, by wiping Nasonov glands with small pieces of filter paper and showed that (contrary to previous supposition) it was not specific for race or colony, so the Nasonov gland is unlikely to be a source of colony odour. More recent work (page 121) has failed to find any difference in the composition of Nasonov pheromone components of different races of honeybees. Foraging experiments are needed in which foragers choose between their own and other colony odours deposited by bees that have not exposed their Nasonov glands.

Bumblebee colonies also have distinctive odours although no food transfer occurs between individual bees. Anaesthetized workers (*Bombus pascuorum* and *B. lucorum*) were attacked when introduced to a strange colony of their own species, but anaesthetized nestmates of the recipient colonies were not. Bumblebee colony odour may be acquired from the nest atmosphere. Individual workers are attacked on return to their own nest after being confined to combs of strong colonies of the same species from which the rightful occupants have been removed, or after being suspended above the combs of strong colonies for one or two hours (Free, 1958).

There remains the possibility that colony odours are genetically determined. This was discounted by Kalmus and Ribbands (1952) because the split halves of a colony produced by a single queen had separately distinct odours after a few days. However, it is now known that a queen receives sperm from several males so her workers are genetically heterogeneous and it is unlikely that offspring from different fathers will be evenly distributed in any artificial colony division. Breed (1983) has provided strong evidence that recognition can be by genetically determined odours. Workers produced from the eggs of different queens were reared in the same colony and kept in the same environment for five days after emergence. When presented to strange bees they were less likely to receive an aggresive reception if they had come from eggs laid in the recipient bees' own colony. If colony odours are (at least in part) genetically determined it would help explain why foragers on return to their hives are accepted as belonging to the colonies concerned, even though each has probably visited only one flower species during its foraging trip and so may be saturated with only one floral odour. Indeed, bees that have collected loads of pollen are readily recognized as alien when introduced to the hive entrance of a strange colony (Butler and Free, 1952).

The possible extent to which colony odour normally has a genetic basis needs reinvestigation. The colonies used should be headed by queens that have been artificially inseminated so all the offspring within a colony are of identical parentage.

If colony odour does prove to be genetically determined individual bees must learn the odours of their half-sisters as well as their own odour shared by their true sisters. Getz and Smith (1983) raised workers of different genetic constitution in the same colony. On emergence they were kept with their full sisters in groups of ten. After five days, bees within a group were more antagonistic to strangers that were half-sisters than to strangers that were full sisters. It is not clear whether this was because the bees innately responded more favourably to their full sisters' odour or whether it was because they had become conditioned to its presence on the other nine workers with which they were confined. The former alternative receives support from an experiment by Getz et al. (1982), who found that within a colony workers of different male parentage segregate non-randomly during colony swarming. They suggested that kin recognition could be responsible for this non-random grouping.

Perhaps colony odour will prove to be a composite of inherited odour produced by various exocrine glands and the adsorbed odour from the hive atmosphere, much of which must derive from the bees themselves. If so, bees must continuously re-learn and re-adapt to their colony odour as it changes.

It would be expected that larvae, pupae and newly emerged bees would also adsorb odour from the environment; yet they are readily accepted without animosity into a strange colony. Hölldobler and Michener (1980) suggested that any colony odour acquired by brood may be masked by the presence of brood-tending pheromones.

Clearly this is a most promising area for investigation, and the information obtained could be used with advantage by beekeepers when uniting colonies or helping to prevent a colony being robbed by others.

Queen recognition

The queen of each colony has a distinctive odour and bees can recognize their own queen from others (Henrikh, 1955). It is essential that guard bees allow their own queen to enter the hive after her mating flight, but repel any others that land at the hive entrance in error. When given a choice a queenless swarm always prefers its own queen to another (Henrikh, 1955; Velthuis and van Es, 1964; Morse, 1972); sometimes the strange queen attracts a small cluster but this usually disappears within a few hours.

Discrimination depends upon the bees being able to make physical contact with the queens (Boch and Morse, 1974). Clustered swarms were given the choice of moving to either their own queen or a strange queen in cages tied to stakes 2 m from the swarm. When the cage walls were made of queen-excluder material (perforated zinc sheets through which the workers but not the queens could pass) 32 of 34 swarms that were tested clustered round their own queen. The workers licked their queen's body and palpated it with their antennae. When the cage walls had 3.2 wires per cm so that the workers could touch the queen but could not enter the cage, swarms still preferred their own queen, but when the cage walls had 32 wires per cm or double walls, so that direct contact with the queen was prevented, no discrimination was evident. Swarms were able to discriminate between recently killed queens of their own and other colonies but not between ethanol extracts of their own and foreign queens.

Feeding colonies for a few days with sucrose syrup strongly scented with either thyme or eucalyptus oils diminished their ability to distinguish between their own and foreign queens (Boch and Morse, 1981). This is somewhat surprising, as adsorption onto the body surfaces of the odour of the sucrose syrup could be expected to enhance the distinctive colony odour and make recognition of their own queen easier – instead it seemed to obstruct discrimination. Boch and Morse (1981) suggested that the workers' chemical receptors became swamped by the scent of the syrup so that they could no longer distinguish familiar from foreign queen odours.

In another experiment, Boch and Morse (1979) divided a colony by a double screen partition and gave each half a queen. Although the hive odours could freely pass though the screen each half was able to recognize its own queen when tested several days later, indicating that the queens had not acquired a similar odour. Therefore, adsorption of odour onto the body surfaces of queens is not alone sufficient to explain their individual odour characteristics.

There is evidence that the individual queen odour is partly genetically determined and so is partly based on pheromones produced by the queens themselves. In laboratory tests Breed (1981) found that small groups of workers accepted queens that were sisters to their own queens more readily than they accepted non-sister queens; furthermore inbred sister queens were more readily accepted than outbred sister queens.

Even when queens are kept in the same environment so that they adsorb the same odours onto their cuticular surfaces, workers can still differentiate between them (Free *et al.*, 1987d). Queens were caged separately, each with a group of 50 workers, in the same incubator and given the same food for ten days. When the workers were then allowed to choose between a chamber containing their own queen and one containing another previously caged queen, most moved into the chamber with their own queen.

In the field (Boch and Morse, 1982) swarms recently dequeened chose to cluster round queens that were sisters to their own queens (produced from a breeder queen that had received semen from a single drone of her own line) in preference to unrelated queens. However, they still preferred their own queen to her sisters, suggesting that the colony odour queens adsorb from the hive atmosphere and food supply are also important in contributing toward recognition.

The workers of a colony may also become conditioned to the blend and quantity of queen pheromone components provided by their own queen. Thus when given the choice, a swarm whose queen has been removed prefers an alien queen that is the most similar in age and reproductive status to their original queen; a swarm previously headed by a mated laying queen prefers an old virgin queen to a young virgin queen (Ambrose *et al.*, 1979).

Mutual recognition by queens

When a colony rears a new queen without swarming the old queen may be killed by the workers immediately, or she can remain alive for several months, and mother and daughter may be found on the same comb without signs of antagonism. In all other circumstances in which more than one queen are present in close proximity, they are hostile to each other.

It is well known that when two virgin queens meet on the comb they will fight until one is killed, and that the first virgin queen to emerge in a queenless colony will sting and kill any others that are still in their cells. It is uncertain whether a queen actively seeks her rivals or whether meeting is by chance; it should be possible to determine in an observation hive the distance from which a virgin queen moves directly toward a queen cell or toward a caged adult queen.

The means by which queens recognize each other has not yet been fully determined. They tend to be more aggresive to each other when of a similar age and physiological condition (Szabo and Smith, 1973). Some of the sensilla on the antennae of queens specialize in 9-ODA perception (Kaissling and Renner, 1968); perhaps queens use their ability to detect 9-ODA to discover any new and potentially hostile queens in their colony, but this presupposes that their sensilla do not become adapted to 9-ODA. However, removal of their mandibular glands does not influence their aggressiveness or fighting ability (Velthuis, 1967; Szabo and Smith, 1973), suggesting that mandibular gland pheromone is not directly involved. Furthermore, queens whose mandibular glands have been removed and whose abdominal tergites have been covered with nail polish do not fight, suggesting that mutual recognition may be by pheromone secreted from the tergite glands (Velthuis, 1967). The effect of covering the tergite glands of otherwise intact queens has yet to be determined.

Nestmate recognition by primitively eusocial bees

Female bees of *Lasioglossum zephyrum* (a primitively eusocial halictid bee) also have individual inherited odours (Barrows *et al.*, 1975). Males looking for mates are able to recognize and respond differentially to these odours. They tend to respond less to a female they have experienced previously. This has a strong biological advantage because such a female would either have repulsed them or have already mated with them.

Members of the small colonies of *Lasioglossum zephyrum*, do not appear to acquire one another's odour but each seems to learn the individual odours of the few bees of the colony.

Guard bees at the nest entrance distinguish their own nestmates from other conspecific bees by odour (Michener *et al.*, 1971; Barrows *et al.*, 1975). Memory of each individual odour lasts for several days. Even after nestmates had been isolated for some days the guards still accepted them but after 12 days of isolation all reintroduced nestmates were rejected (Barrows *et al.*, 1975).

The acceptance of non-nestmates increases with their degree of relationship to the guards' own nestmates and hence with the familiarity of their odours (Greenberg, 1979; Buckle and Greenberg 1981). However, adults less than 48 hours old are generally accepted, probably because they have not yet acquired distinctive odours and as they do so, these are learned by their nestmates.

Chapter twelve

TRAIL AND DETERRENT PHEROMONES

Introduction

There is strong evidence that honeybee workers involuntarily deposit pheromones that attract others both when foraging and when entering their nest. Because the same pheromone appears to be concerned for each activity I have chosen to call it by the commonly accepted name of 'trail' pheromone rather than the previous names of 'footprint' pheromone (Butler *et al.*, 1969) and 'forage-marking' pheromone (Ferguson and Free, 1979). It has not been identified.

It also appears probable that foraging honeybees are able to mark flowers whose nectar sources have been depleted – helping to avoid unproductive visits.

Trail pheromone at the nest

Butler *et al.* (1969) showed that workers entering their hives deposit a persistent attractive material. Glass entrance tunnels that have been marked with this trail pheromone are much preferred by homecoming bees to clean glass entrance tunnels. The attractiveness of an entrance tunnel increased with the number of workers that had previously used it up to about 400 workers, thereafter its attractiveness failed to increase further. The trail pheromone of workers from another colony was also attractive (page 101), but slightly less so than that of workers from a bee's own colony. The trail odour probably is often used to help bees orient to a change in the location of the entrance to their hive (Ribbands and Speirs, 1953; Lecomte, 1956; Butler *et al.*, 1970), especially when the bees habitually land at the site of the old entrance and walk across the outside of the hive to the new one. No attempts appear to have been made to find whether such a trail is also unidirectional. The odour of trail pheromone deposited on inert material such as glass or nylon mesh induces temporarily disorientated bees to expose their Nasonov glands (page 114) at the hive entrance (Ferguson and Free, 1981).

The floor and the inside walls of the hive or nest and the combs themselves are also probably marked with trail pheromone. The accumulation of trail pheromone on comb may partially explain why old comb is more attractive than new (page 81). Butler (1967) found that bees he had trained to forage in a darkened arena produced an odour trail between their hive and the dish of sucrose syrup.

Trail pheromone while foraging

It is well established that a glass dish on which bees have been foraging for sucrose syrup is more attractive to potential foragers than a clean dish, probably because of an attractive trail pheromone the foraging bees have left behind.

Experiments that have demonstrated this involved training bees to collect sucrose syrup from tubes or dishes placed on a circular table. The tubes or dishes with syrup were replaced by empty ones, provided with different odours and placed equidistant from the table centre; the number of bees that landed on or touched each tube or dish was compared. The table was rotated continuously so the bees did not become conditioned to any particular position (see Ribbands, 1954; Butler *et al.*, 1969; Ferguson and Free, 1979).

Bees visiting a site mark it with an attractive pheromone irrespective of whether they have foraged successfully there. Ribbands (1954) showed that it was only necessary for a bee to land briefly on a particular empty tube for it to prefer that tube subsequently, Free (1970b) found that would-be foragers were attracted to the odour bees had left on a glass sheet covering model flowers from which they could not obtain food, and Ferguson and Free (1979) demonstrated that dishes on which bees had landed and had not foraged became attractive to others. It has been shown that small blocks of plaster of Paris, small sheets of glass or squares of wire gauze that had been kept on the floor of a hive just inside the entrance for a few hours and on which numerous bees had walked, became attractive to foraging bees and retained this attractiveness for many hours (Butler *et al.*, 1969, Ferguson and Free, 1979). Thus it appears unlikely that the attractive pheromone involuntarily deposited by foragers is exclusive to foraging, but is probably the same general trail pheromone responded to at the hive entrance. It is certainly a more primitive form of communication than the Nasonov pheromone.

It seems that foraging bees may also have a preference for trail odour deposited by bees of their own colony. Bees from two colonies were trained to two separate but adjacent dishes of dilute sucrose syrup; the dilute syrup in each dish was then replaced by concentrated syrup so that dancing and recruiting were encouraged. Newcomers that arrived were preferentially attracted to the dish visited by their nestmates (von Frisch and Rösch, 1926; Kalmus and Ribbands, 1952), and so deposition of trail pheromone at a source of forage favours survival of their own colony. Undoubtedly the bees were exposing their Nasonov glands as they arrived at the dishes, but there is no evidence that Nasonov pheromone is colony specific (page 103).

There is no information on the chemical identity of the trail pheromone. It has been shown that bees are attracted to an ethanol extract of glass beads on which bees have stood to forage (Lecomte, 1957) and a dichloromethane extract of the deposit left by bees that gathered on empty dishes (Ferguson and Free, 1979).

Chauvin (1962) thought that trail pheromone is probably produced by the Arnhart's gland located on the tarsal extremities but it now seems that the trail pheromone is widely distributed on a bee's body. The odour of head, thorax or abdomen alone is attractive and will elicit an alighting response of bees searching for food, although the thorax and dorsal surface of the abdomen are especially effective. (Butler et al., 1969; Ferguson and Free, 1979). The increased response to the dorsal surface of the abdomen possibly reflects the presence there of Nasonov pheromone. Irrespective of its origin, the trail pheromone must move to the feet from whence it is deposited, as other parts of a walking bee's body (with the possible exception of the tip of the abdomen (Butler, 1969)) rarely touch the substrate. Because bees leave a trail odour on glass it is likely they do so even more readily when trampling over the petals of a flower. When foraging on flowers all parts of a bee's body often rub against the petals, stamens and stigma so a rapid rate of deposition would be possible.

Free and Williams (1974) found that wire gauze on which A. florea foragers had stood to reach sucrose syrup below induced other foragers to alight on it, so it is likely that A. florea workers also produce an attractive trail pheromone. It is probably of special significance in the communication of rewarding floral sources to other A. florea foragers as, unlike A. mellifera, they do not release Nasonov pheromone (page 118) in such circumstances.

Although the trail pheromone is very effective in releasing alighting responses of bees conditioned to a source of forage, and by bees recruited to the area, it is not known at what distance it is perceived and responded to. If foragers are attracted from a distance, it would be worth exploring the possibility of using a synthetic preparation of the pheromone to apply to crops needing insect pollination.

Trail pheromones of stingless bees

Various stingless bees (Meliponinae: *Trigona postica*, *T. ruficrus* and *T.*

geotrigona) do not make orientated communication dances, but deposit scent marks every few metres between the nest and food to form a trail that alerted nestmates follow (Lindauer, 1956; Lindauer and Kerr, 1958; Nedel, 1960). When a scout bee has discovered a food source it usually makes several trips between its nest and food before it lays down a pheromone trail. Scent marks are deposited on leaves, branches, pebbles and even clumps of earth. Distances between adjacent scent marks vary according to the species concerned from 2 m or less (*T. bipunctata*) to between 10 and 30 m (*T. trinidadensis*) (Kerr *et al.*, 1963).

Workers of *T. subterranea* place their scent marks in an irregular manner and the distance separating adjacent marks is very variable. The first mark is placed at the food source and the second only 30–50 cm from it. Thereafter marks are placed 1–5 m apart in the direction of the nest. Individual bees have their own characteristic marking patterns (Blum *et al.*, 1970). Different species also have their own preferred height at which to deposit scent trails (Kerr *et al.*, 1981).

Trail odours last only 8–19 minutes without reinforcement (Kerr *et al.*, 1963). However, it is possible that the odour released by *T. cupira* is strong enough to form an aerial odour trail in the calm conditions of a tropical forest (Kerr, 1969). While foraging on flowers workers of *T. spinipes* produce a strong odour from their mandibular glands that may attract recruits (Kerr, 1973).

The scent trails of some species are more effective than honeybee dances in assembling recruits and in addition can provide information on the vertical component (Lindauer and Kerr, 1958).

However the scent trail alone of other species appears to be insufficient to elicit trail-following behaviour of recruits who need to be guided by the scout bee to the food on their initial trip, although thereafter they are able to follow the trail on their own (Cruz-Landim and Rodriquez, 1967).

The trail pheromone is secreted by the mandibular glands of bees that have reached 40–50 days old; the mandibular glands of younger bees are insufficiently developed (Cruz-Landim and Ferreira 1968). Individuals tend to respond to the trail of their own species only but limited interspecific trail following does occur. Thus *T. xanthotricha* can follow trails of *T. positca* but not *vice versa* (Kerr *et al.*, 1963); probably these two species have only minor differences in chemical components of the trail pheromone.

(E)-and (Z)-citral are the dominant components of the mandibular gland secretion of *T. subterranea* (Blum *et al.*, 1970). Bees flying toward a food source examined small wooden blocks treated with citral that had been placed *en route*, and landed near those that had been placed at the food source itself. However, when the citral-treated blocks were placed inside a *T. subterranea* nest they were immediately attacked; the agitated workers made no attempt to leave the nest. Hence, the same chemical releases both trail following and alarm, and the mandibular gland secretion possibly serves both functions.

The mandibular gland secretion of *T. spinipes* which produces well-defined trails (Kerr *et al.*, 1963) contains 2-heptanol as the major volatile component, and 2-nonanol, 2-tridecanol, octyl octanoate and octyl decanoate as minor components (Kerr *et al.*, 1981); 2-heptanol alone was successfully used to produce artificial trails followed by *T. spinipes* workers.

Scent marking of unproductive food sources

Núñez (1967) produced evidence that honeybees may mark unproductive food sources with scent. He trained bees to stand on the bottom rims of 12 vertical glass outer tubes while collecting sugar syrup from a capillary tube inside each. When some of the capillary tubes contained no sugar syrup bees continued to alight on the outer tubes, but only briefly. When the positions of the outer tubes were exchanged so that outer tubes which had previously housed capillary tubes with syrup now housed empty tubes, they were still subjected to prolonged visits; and the bees still landed only momentarily on outer tubes which had enclosed empty capillary tubes, although they now enclosed full ones.

Driving a current of air over the ends of the outer tubes in an effort to disperse any odour present resulted in both types of outer tube receiving prolonged visits and very few transient visits, which suggested that it was a repellent odour deposited at an unproductive food source which discouraged foragers rather than an attractive odour at the productive food source which encouraged them. Perhaps, however, the current of air removed the more volatile repellent pheromone but not a less volatile attractive one.

Free and Williams (1983) provided evidence for the existence of both attractive and deterrent pheromones. They used 'artificial flowers', each consisting of a small conical flask with a plastic platform resting on top of it on which the bees could land. A glass capillary tube 50 mm long extended from a hole through the centre of the plate into the flask. A continuous supply of forage could be provided by allowing sucrose syrup from the flask to rise up the tube to replace any removed by foragers. The supply could be terminated by allowing air to enter the tube. Bees trained to forage at the artificial flowers were presented with a mixture of flowers that had always provided sucrose syrup (rewarding flowers), that had never provided sucrose syrup (unrewarding flowers), that had their supply of sucrose syrup terminated ten minutes previously (terminated flowers) and that had never been presented before (clean flowers). More bees visited rewarding flowers than unrewarding or terminated, and terminated flowers were visited more than unrewarding flowers. All of these findings could reflect the deposition of attractive pheromone on flowers providing forage. However, clean flowers received fewer visits than rewarding or terminated but more than unrewarding, indicating that the unrewarding were marked with a deterrent pheromone. The results were similar whether the landing platform was plastic, covered with filter paper or a flower petal, or whether or not a synthetic scent was present, so it appeared likely that both attractive and repellent pheromones may be used to signal the presence or absence of forage in natural flowers. There is now evidence that natural flowers that have been recently visited are less favoured by foraging bees (Corbet *et al.*, 1984).

It seems that sources from which bees are able to obtain food are marked with an attractive odour, probably the stable 'trail' pheromone referred to above. Probably this occurs inadvertently. However, even though bees land on unrewarding flowers (and therefore mark them inadvertently with trail odour) the deterrent pheromone deposited results in an overall repellent signal. It seems probable that bees may mark in the same way flowers they themselves have depleted (Corbet *et al.*, 1984), but it has yet to be demonstrated that a bee which has depleted an artificial flower of sucrose syrup then proceeds to mark it and deter further bee visits.

Deterrence from unprofitable flowers should be short-term only while they are replenishing their stocks of nectar; the biological efficiency of the repellent pheromone would be reduced if it persisted for longer then necessary. It is likely that a relatively volatile pheromone is used but the source of such a pheromone is unknown. Possibly 2-heptanone secreted by the mandibular glands is responsible; it is known to repel foragers from food sources (e.g. Simpson, 1966; Butler, 1966b; Ferguson and Free, 1979; page 144).

Pheromone signals used to mark flowers as attractive sources of forage should be longer-term, as there may be no potential foragers in the vicinity for several minutes or even for some hours. As Corbet *et al.* (1984) point out, the ultimate message given by the pheromones left on a flower will depend on the balance of the attractive and deterrent pheromones. For some minutes after a flower has been visited and depleted of nectar the deterrent pheromone with which it has been marked will have a dominant influence. Thereafter, as it volatilizes, the influence of the more persistent attractive signal left by foragers which were obtaining food will reassert itself and the flower will again be attractive to foragers, by which time the supply of nectar will be replenished.

Chapter thirteen

ATTRACTION:
NASONOV PHEROMONE

Introduction

The Nasonov scent gland of the worker honeybee lies on the dorsal surface of the 7th abdominal tergum (Fig. 2.3) and consists of a mass of large glandular cells which secrete pheromone through 600 or so minute ducts into a groove or canal between the 6th and 7th terga (McIndoo, 1914; Snodgrass, 1956). The Nasonov gland and canal is usually concealed by the overlapping part of 6th tergum and the bee exposes them by flexing the tip of its abdomen downward. During exposure the bee usually stands with its abdomen elevated and fans its wings, drawing air over the exposed gland and canal so facilitates dispersal of the Nasonov pheromone. The gland is absent in queens and drones.

The Nasonov gland is present in workers of *A. cerana*, *A. dorsata* and *A. florea* as well as in *A. mellifera*, but unless otherwise stated all of the following information applies to *A. mellifera*.

The Nasonov gland is named after the Russian who, in 1883, first described it; but the function of the gland was then unknown although it was supposed to help in secreting surplus water, possibly from nectar the bees had collected (Zoubareff, 1883).

This erroneous concept was corrected by Sladen (1901, 1902, 1905) who realized that the odour from the gland attracted other bees. This function was confirmed by experiments demonstrating that groups of 50 queenless bees were attracted to cluster on small cages containing excised Nasonov glands and that individual bees preferred to move toward excised Nasonov glands in a two-choice olfactometer (Free and Butler, 1955). Nasonov pheromone adsorbed onto filter paper is also effective in an olfactometer (Boch and Shearer, 1962). Bees in the olfactometer were equally attracted to excised

Nasonov glands of members of their own colony and to those of workers from strange colonies.

It is now known that Nasonov exposure occurs in a variety of circumstances in which aggregation of workers occurs and so the one pheromone may perform multiple functions. Honeybee workers release pheromone from their Nasonov glands in three main behavioural contexts:

1 when in a swarm that is moving or forming a cluster (e.g. Sladen, 1905; Jacobs, 1924; Velthuis and van Es, 1964; Morse and Boch, 1971);
2 when marking the entrance to their nest (e.g. Sladen, 1901; Ribbands and Speirs, 1953)
3 when indicating a source of water on which they are foraging (e.g. Free and Williams, 1970).

The various other circumstances in which pheromone release occurs are probably each a derivation of one of these, and are sometimes adaptations to artificial conditions created by the beekeeper. It seems that, in general, the responding bees are those that are disorientated or need aid in orientation.

It was supposed that individual colony odours (page 103) arose at least partly from the Nasonov gland (Jacobs, 1924; Kalmus and Ribbands, 1952) but it is now known that this concept is erroneous and the Nasonov pheromone is neither race nor colony specific (Renner, 1960).

Nasonov release during colony movement

Extensive release of Nasonov pheromone occurs during different stages of the

A worker (*Apis mellifera*) fanning and exposing its Nasonov gland at its hive entrance

transit of a swarm from its parent colony to its new nest site (e.g. Sladen, 1905; Morse and Boch, 1971).

After leaving its parent colony the swarm clusters nearby on a tree branch or other support. The first bees to settle expose their Nasonov glands, and the released Nasonov pheromone is dispersed by fanning with the wings; this attracts other bees still airborne to the site. Sladen (1905) removed the queen from a swarm and, about 15 minutes later, placed her a few yards away; as soon as the bees discovered her they exposed their Nasonov glands, fanned and clustered around her.

Once the swarm has clustered, scout bees search for a permanent nest site. Sometimes they may locate potential nest sites before the swarm leaves its parent colony (Lindauer, 1957). Scout bees which have discovered a nest site return to the cluster and perform direction- and distance-indicating dances. They then go back to the new nest site, release Nasonov pheromone near its entrance, and so attract more scouts to investigate the site.

When most of the scout bees agree that a particular nest site is suitable the swarm leaves its temporary clustering site and flies to the new nest. Scout bees may lead the swarm and release Nasonov pheromone en route (page 27), but this is not certain (Lindauer, 1957; Avitabile et al., 1975).

Some scouts are already at the nest site releasing Nasonov pheromone while the swarm is airborne, Nasonov pheromone is released by most bees in the swarm as they land at the nest entrance and while walking into the nest. Nasonov gland exposure also occurs in certain artificial circumstances – for example, when a beekeeper introduces a swarm or colony to an empty hive and a broad stream of bees moves slowly into the hive entrance, a mass release and dispersal of Nasonov pheromone occurs and so attracts stragglers.

Nasonov release during reorientation

Even after a colony is well established in its nest or hive, Nasonov pheromone may act as an orientation aid. It is often released by young bees after their first orientation flights (Hazelhoff, 1941), but any disturbance that causes flying bees to become disorientated often causes them to expose their Nasonov glands and fan when eventually they alight near the nest or hive entrance. Every now and then they walk a short distance towards the entrance before halting and continuing to release scent and fan; the bees continue these activities until actually inside the nest (Sladen, 1901). However, not all the bees behave in this fashion; even after the entrance has been closed for several seconds previously some bees walk into the hive, apparently without hesitation and without scenting. Perhaps differences in the amount of pheromone present, age and occupation of the bee and difficulty in orientation decide the threshold of the scenting response.

Attempts have been made to determine the factors important in eliciting

Nasonov gland exposure by such disorientated foragers (Ferguson and Free, 1981). An established honeybee colony always contains wax comb, food stores, propolis, adult workers and a queen, and odours from each of these facilitate nest recognition and induce Nasonov gland exposure. Brood is absent during part of the winter and other dearth periods; its odour (page 70) does not influence Nasonov pheromone release. Recently killed drones and workers also fail to do so.

Whereas the 9-HDA component of a queen's mandibular glands releases Nasonov exposure the 9-ODA component does not. In contrast, the presence of 9-ODA induces queenless workers to cluster, whereas the addition of 9-HDA lessens the effect (page 26; Ferguson et al., 1979). This provides an example of different components of a pheromone being differentially effective in different behavioural contexts.

Sladen (1901, 1902, 1905) was the first to suggest that the odour dispersed by fanning and scenting bees attracted other members of their colony, and indeed he may be regarded as the pioneer of studies of pheromones of bee communities.

Ribbands and Speirs (1953) changed the position of the entrances of hives and demonstrated that the likelihood of returning bees exposing their Nasonov glands increased with the difficulties of reorientation; furthermore they showed that the Nasonov odour produced by temporarily disorientated worker bees guided other bees that were still disorientated to the safety of the hive. Other experiments have confirmed the attractiveness of Nasonov pheromone (page 14).

Because bees at the nest (or hive) entrance that are fanning and releasing Nasonov scent are faced toward the entrance and so fan the pheromone away from the nest, it is likely that the signal from the Nasonov gland is mixed with odours emanating from the nest itself. Undoubtedly this must enhance the location of the entrance by lost individuals.

As Sladen (1901) pointed out 'calling' is infectious and he suggested that scenting by the first workers of a colony to discover their hive appears to rapidly release such behaviour by their companions. Ferguson and Free (1981) confirmed this was so (page 125). Synthetic Nasonov pheromone itself induced exposure of the Nasonov gland; this helps explain the rapid increase in the proportion of scenting bees that occurs as bees cluster, as a swarm moves into a new nest, or as foragers enter their hive after being temporarily denied access, all circumstances in which bees aggregate.

When a virgin queen is about to leave her hive to mate, many of her workers assemble at the hive entrance and release Nasonov pheromone (Ruttner, 1956); this probably serves the double purpose of guiding the virgin queen to the entrance and helping her to find the hive on her return. It is well known among beekeepers that if a colony is examined about 24 hours after its queen has been removed, many of the worker bees release Nasonov

pheromone and fan, although the exact circumstances initiating this behaviour by queenless bees has yet to be determined. It may be an adaptation
of natural behaviour in which a queen is guided back to her colony after a
mating flight.

Nasonov exposure also occurs in various artificial circumstances connected
with reorientation; for example when small groups of workers are confined to
cages with or without a queen they periodically scent and fan in unison. The
precise conditions releasing such scenting behaviour are not clear, although
they presumably in some way reflect conditions under which it occurs
naturally. When bees are released from captivity in such small cages, many
expose their Nasonov glands at the cage entrance before taking flight (Butler
and Free, 1952). The presence of comb in the cages encourages such
behaviour. After the bees have taken flight the presence or absence of comb
in cages does not influence the bees' return to them, but more bees return to
cages at which scenting has occurred (Free and Williams, 1971).

Nasonov release during foraging

Bees also expose their Nasonov glands when foraging, but without fanning so
the pheromone release is non-directional. Nasonov exposure by *A. mellifera*
foragers on flowers has been reported only twice. von Frisch and Rösch
(1925) found that bees foraging in a glasshouse for visible drops of nectar
produced by cut flowers of *Robinia viscosa*, and pollen produced by *Rosa
moschanta*, danced when they returned to their hive, and scented on their
next visit. Free and Racey (1966) saw a bee expose its Nasonov glands when
collecting the abundant nectar from *Freesia refracta* flowers in a glasshouse.
The extent to which workers of other *Apis* species release Nasonov pheromone when visiting flowers needs investigation; Butler (1954c) observed that
both *Apis cerana* and *Apis florea* did so in Sri Lanka but Free and Williams
(1979) did not see *A. florea* do so in Oman.

Although *A. mellifera* foragers rarely, if ever, expose their Nasonov glands
on flowers in the open they readily do so when collecting from a favourable
supply of sugar syrup. von Frisch (1923) found that when he sealed together
with shellac the rear abdominal terga of honeybees collecting concentrated
sugar syrup so that they could no longer expose their Nasonov glands, they
attracted fewer recruits than previously and no more than those foraging on
syrup of a low concentration.

The effectiveness of Nasonov pheromone in attracting foragers was also
demonstrated by Renner (1960). In a series of experiments more than twelve
times as many recruits searching for sugar syrup landed at dishes scented
with Nasonov pheromone (wiped onto filter paper) as landed at unscented
dishes. Free (1968a) showed that scout bees, or searching bees recruited by
bee dances, are attracted to the odour of excised Nasonov glands at dishes
containing sugar syrup and this induces them to land.

Free (1968a) found that whereas some bees that released scent had exposed their Nasonov glands before they landed at a dish and continued to expose them for a few seconds afterwards, others exposed their glands only momentarily, usually soon after alighting. The tendency to scent differed greatly between individuals, some not scenting at all during numerous consecutive trips, and some consistently doing so more readily than others. The impression was obtained that the bees scenting most consistently tended not only to scent while approaching a dish, but also to scent for longer than the others while feeding.

The precise factors associated with the forage that lead to scenting need to be determined. The sugar content of the syrup, an increase in its sugar content, and the ease with which it can be imbibed seem to be important (Frisch, 1967; Pflumm, 1969; Pflumm et al., 1978).

Bees usually do not scent until after the first few visits to a source of sugar syrup (Free, 1968a). Such a delay would presumably have the biological advantage of not guiding other bees to a very transient food source or one that is toxic, and would help ensure that there is an absence of predators in the area concerned. Once a bee begins to scent it usually continues to do so on subsequent trips during the day.

Bees that dance on their return home from collecting syrup are more likely to scent on a subsequent trip than those that do not dance (Free and Williams, 1972) and both scenting and dancing are stimulated by an increase in the attractiveness of the forage (Free, 1968a). Data collected by von Frisch (1923) indicates that the longer syrup foragers dance, the longer they circle in the air above the dish – perhaps while circling they are exposing their Nasonov glands. However, dancing and scenting are not closely correlated (Free and Williams, 1972) and each may therefore be released by a different stimulus or perhaps by different intensities of the same stimulus. Thus, although the presence of a floral odour at a food source encourages dancing (Lindauer, 1948) it discourages scenting by bees collecting syrup (Free, 1968a; Wenner et al., 1969) or water (Free and Williams, 1970). However, it makes no difference to the duration of any scenting that does occur (Wilhelm and Pflumm, 1983). Whereas the presence of strong floral scents makes bees less likely to add their own, it seems that bees are particularly likely to add their own Nasonov scent to a food source that lacks one, and so facilitate its discovery by others.

In contrast to the behaviour of bees that are clustering, the presence of Nasonov gland odour fails to influence Nasonov odour release by foragers at a food source (Free, 1968a; Pflumm and Wilhelm, 1982). This is not surprising. If Nasonov odour encouraged scenting the apparent attractiveness of a food supply could be too rapidly magnified and the accidental release of scent could start a false communication: if the odour discouraged scenting it would defeat its own object.

Odourless food sources are rare in nature; however, clean odourless water is frequently more easily obtainable by bees than water containing the odour of decaying vegetable and animal material. It has been found that when worker bees have made several consecutive trips for water many of them expose their Nasonov glands at the source of supply and especially when it lacks an odour of its own (Free and Williams, 1970). Although the bees were not found to expose their Nasonov glands as readily or as consistently when collecting water as when collecting concentrated sugar syrup, this may have merely reflected the syrup's relative attractiveness at the time, and perhaps when water is in greater demand scenting is more pronounced. In contrast to Nasonov release at the hive entrance (Ribbands and Speirs, 1953), that at a water site was not related to any difficulty bees had in locating it. Probably under natural conditions, Nasonov pheromone released by foraging bees is usually associated with water collection and not with nectar or pollen collection. This would also help explain why fanning is not associated with Nasonov release in these circumstances.

Bees recruited to an attractive natural food source do not follow the dance directions very precisely (von Frisch, 1967) nor is it desirable that they should do so. A natural food crop is usually distributed over a considerable area, and, if the arrival points of the recruits are also widely distributed, a colony can exploit it more efficiently than if all its foragers concentrated on a small area. However, most naturally occurring water sites, other than early morning dew or rain, are more discrete than those of nectar and pollen and have less pronounced odours, and the exposure of the Nasonov glands by foragers at such sites helps recruits to find them sooner.

It is not known whether bees expose their Nasonov glands when foraging on dew or raindrops, but it is doubtful whether this would commonly occur as they are likely to evaporate before the bees make sufficient trips to begin scenting.

The exposure of Nasonov glands by bees foraging for nectar in glasshouses or on dishes of odourless sugar syrup is probably an adaptation of the behaviour that naturally occurs when collecting water. Perhaps dancing and Nasonov gland exposure are more closely correlated when *A. mellifera* foragers are collecting water; this would be interesting to discover.

All instances of Nasonov release cited so far have occurred outside the nest itself, but it may also occur within the nest. I have seen workers in a glass-walled hive expose their Nasonov glands when a sudden decrease in temperature has caused them to move together into a compact cluster.

Perhaps Nasonov pheromone is used to mark routes within the nest and particularly to food stores remote from the cluster. A path of fanning bees has been observed between the combs and a feeder containing sugar syrup inside the top of a hive (Jacobs, 1924). This behaviour was further studied by Levchenko and Shalimov (1975) who showed that the tendency of bees to scent increased with the distance between the combs and supply of sugar syrup. Nearly all the bees that scented did so while on the final part of their approach to the feeder; this seems anologous to Nasonov exposure by flying bees prior to landing at a source of water or food.

Chemical composition

Identification of the chemical components of the Nasonov pheromone has been a slow process and may not yet be completed. Those so far identified are given in Table 13.1 and Fig. 13.1.

Table 13.1   Identification of components of Nasonov pheromone.

Component		Identified by
Terpenic alcohols	Geraniol	Boch and Shearer 1962
	Nerol	Pickett *et al.*, 1980
	(E,E)-Farnesol	Pickett *et al.*, 1980
Terpenic aldehydes	(E)-Citral	Weaver *et al.*, 1964
	(Z)-Citral	Weaver *et al.*, 1964
Terpenic acids	Geranic acid	Boch and Shearer 1964
	Nerolic acid	Boch and Shearer 1964

The tentative identification by Weaver *et al.* (1964) of citral as a minor component of the Nasonov pheromone was confirmed by Shearer and Boch (1966). However, they were unable to detect it in Nasonov secretion directly it was collected, and did not find it was formed in secretion that was immediately cooled in dry ice. But at room temperature, the amount of citral slowly increased for the first 15–20 h after secretion. Because this was associated with a decrease in the amount of geraniol even in a nitrogen atmosphere, they suggested that citral was formed through the oxidation of geraniol and that direct oxidation by air is unlikely. Blum (1971) suggested that the transformation might be by enzymic oxidation, and Pickett *et al.* (1981) presented evidence for the existence of a highly specific enzyme system that converts geraniol into (E)-citral and enables the bee to control the ratio of the two components. This helps to explain why (E)-citral can be one of the three most active components of the Nasonov pheromone (page 123), despite being one of the least abundant and most volatile.

Because of differences in the volatility of the components, their relative proportions will change with distance from the gland. Whereas the proportion of (E)-citral to geraniol in the Nasonov secretion was 1.3:100 immediately after release, it was 5:100 in pheromone taken from the air immediately above the gland (Pickett *et al.*, 1981). Perhaps a bee that is orientating to Nasonov pheromone follows a gradient of component concentrations. We have no information on the distance from which a bee can perceive Nasonov pheromone and align its movement accordingly.

Sladen (1902) found that only some of the Nasonov glands he excised produced scent and Jacobs (1924) noticed that little pheromone is secreted in

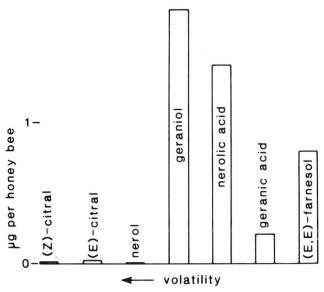

Figure 13.1 Proportions of different components present in Nasonov pheromone (after Pickett *et al.*, 1980).

winter. Boch and Shearer (1963) found that, on average, the amount of secretion present increased with the age of the bees. Newly emerged bees had no geraniol; there was about 0.3 µg per bee when they were guards (10–17 days) and about 1.0 µg per bee (maximum 1.5 µg) when they began to forage (18–20 days). In spring, as the colony population grew, bees had more geraniol than in the winter. These results were confirmed by Pickett *et al.*, 1981; however, whereas they too found that in March and April the amount of geraniol increased, the amount of (E,E)-farnesol remained the same as in January, and also (unlike geraniol) the amount of farnesol did not steadily increase with age after the first week of life. This suggests that the farnesol component may have a different function to the geraniol component; this is supported by experimental work (pages 125, 126 and 127).

Changes in pheromone amount appear to be associated with structural changes in the glandular cells (Vecchi, 1960). It would be interesting to compare the onset of Nasonov gland exposure in the life of a bee with the quantity and composition of Nasonov pheromone present.

No studies have been made of the composition of the Nasonov pheromone of *Apis* species other than *Apis mellifera*.

Electroantennogram responses

Electroantennography (EAG) has been used to study the responses of worker honeybee antennae to Nasonov pheromone and its components. Kaissling and Renner (1968) showed that one type of olfactory receptor cell, which is present in abundance in the antennae of queen, worker and drone honeybees, responds to the Nasonov pheromone. Beetsma and Schoonhoven (1966) and Vareschi (1971) showed that worker antennae responded to geraniol, nerol and citral. Williams *et al.* (1982) obtained antennal responses to all seven known components. There was a greater response to E isomers (i.e. E-citral, geraniol, geranic acid and (E,E,)-farnesol) than to Z isomers (i.e. Z-citral, nerol and nerolic acid; and all Z isomers received a lower response than any of the E isomers. However, the rank order of response to the E isomers differed with the concentration of the components. Thus (E)-citral gave the largest EAG response at 10 μg source concentration but the lowest at 1 μg concentration, while geraniol gave the lowest at 10 μg concentration and the second highest at 1 μg concentration. Thus it seems that although electroantennography may be useful to show that a honeybee has the ability to recognize a component, the extent of the response is extremely variable and does not appear to be closely related to the response of the bees in natural circumstances – as indicated in the following sections.

Behavioural bioassays

Bioassays have been developed to test the role of Nasonov pheromone components in the three main behavioural contexts in which bees release the pheromone, i.e. when foraging, when marking the entrance to their nest, and when in a swarm that is forming a cluster.

Bioassay: cluster formation

A bioassay used to test the effectiveness of different components in attracting clustering bees employed a roundabout with arms 50 cm long that rotated at one revolution per minute. It was suspended from the centre of a large bee-proof nylon cage. Small cylindrical double-walled cages of perforated zinc,

Table 13.2   Preferences shown by clustering bees for Nasonov pheromone components (after Free *et al.*, 1981a).

Choice order for individual components	Choice order for mixtures less the following individual components
Nerolic acid	Geranic acid
Geranic acid	Nerol
(E)-Citral	(Z)-Citral
(Z)-Citral	(E,E)-Farnesol
Geraniol	(E)-Citral
Nerol	Geraniol
(E,E)-Farnesol	Nerolic acid

Bioassay: Honeybee (*Apis mellifera*) worker attracted to cage containing Nasonov pheromone component

each containing a porous polyethylene block dosed with a chemical under test, were suspended from the ends of the roundabout arms, about one metre above the ground. Between 1500 and 2000 worker bees, taken from their colony 30 minutes earlier, were deposited on the floor of the nylon cage beneath the roundabout. The small cylindrical cages they chose to cluster on, and hence the chemical components preferred, were recorded (Ferguson *et al.*, 1979).

Bioassay: Bees clustered on cage containing synthetic pheromone

The components most important in inducing clustering were nerolic acid, (E)-citral and geraniol; removal of any of these three from a mixture of components greatly reduced attractiveness (Table 13.2). (Z)-Citral and geranic acid were also attractive. The commercial preparation of (E)-citral which contains (Z)-citral was preferred to either component in the purified state. Whereas purified nerolic acid was preferred to the more readily obtainable commercial preparation containing geranic acid, the latter stimulates bees to expose their own Nasonov glands (page 126) and its presence with nerolic acid may achieve a net positive effect (Free *et al.*, 1981a, 1982a, 1984b).

A mixture of all seven components of the Nasonov pheromone in equal proportions was at least as attractive (when pure components were used) as, or more so (when unpurified components were used) than, a mixture of all components in natural proportions. Because the amounts of two of the three most important components (geranic acid and geraniol) were similar in each, the enhanced attractiveness of the former mixture probably arose from the greatly increased amount of the third important component, (E)-citral. The enzymic mechanism for regulating Nasonov pheromone composition during pheromone release (Pickett *et al.*, 1980) which involves conversion of geraniol into (E)-citral may be imitated in synthetic mixtures containing an artificially high proportion of (E)-citral. Alternatively, increasing the amount of (E)–citral (which is one of the three most active ingredients) may result in a heightened sensory response, as would be the expected outcome of increasing the amount of total pheromone.

Bees were reluctant to cluster on the roundabout when only nerol or (E,E)-farnesol was present. Removal of these components from a mixture of all pure components of the Nasonov pheromone did not affect its attractiveness, and their removal from a mixture of impure components actually increased its attractiveness.

As a result of these experiments a lure was developed (page 129) that was based only on unpurified components and without nerol and (E,E)-farnesol.

Preliminary tests with *A. dorsata* in Malyasia indicate a preference for four of the *A. mellifera* Nasonov components in the following order: citral, nerol, geraniol and farnesol (Ramli *et al.*, 1984). However, there is no information on the chemical composition of *Apis* species other than *A. mellifera*.

Bioassay: marking nest entrance

A bioassay has been devised in which bees attempting to return to their hive enter an enclosed tunnel of clear plastic in which they are subjected to an air current bearing the odour of single Nasonov components, or mixures of Nasonov components from which one has been removed. The effect of the odours on releasing Nasonov gland exposure is then recorded (Ferguson and Free, 1981; Free *et al.*, 1983b).

When tested singly (E)-citral, (Z)-citral, geraniol, nerolic acid and geranic acid induced the bees to release Nasonov pheromone in the tunnel, and mixtures lacking only one of these components stimulated pheromone release less effecively. In contrast to the other components, (E,E)-farnesol and nerol did not encourage Nasonov exposure when tested singly, and omission of either from the Nasonov mixture increased the attractiveness of those that remained.

Bioassay: foraging

The foraging bioassays for Nasonov pheromone are of three basic types that test the reactions of scout bees, recruits and foragers. In the first type of bioassay petri dishes without food, but with either no odour or the odour of one or more pheromone components, are placed alternatively in a circle, or in a Latin Square arrangement, on a lawn or other area devoid of obvious orientation marks, and the scout bees that investigate each are counted (e.g. Free, 1962).   In the second type of bioassay foragers are trained to a petri dish and, while bees are still foraging on it, other petri dishes with different treatments are placed nearer to the hive and the number of recruits that land on them are counted (e.g. Free, 1962; Boch and Shearer, 1964, 1966; Weaver et al., 1964).

In the third type of bioassay foragers are trained to collect sugar syrup from a petri dish which is then removed and replaced by empty dishes with two or more different treatments, and the foragers that land on each are counted (e.g. Free, 1962; Renner, 1960).

Because foraging bees approach an odour source from downwind, the location of the dishes, with different odour treatments, in relation to the wind direction is most important. If two dishes containing sucrose syrup are located 500 mm apart in line with the wind direction each will receive an approximately equal number of foragers, although the upwind dish may be slightly favoured. When the two dishes are empty, bees landing at the downwind dish tend to take flight again and land at the upwind dish; those that take flight from the upwind dish or overshoot it turn into the wind again and land at the upwind dish but rarely at the downwind dish, so the upwind dish accumulates most bees. Provided that bees have previously been trained to collect sucrose syrup from the dishes the same result is achieved even when the dishes are unscented (Free, 1981). Therefore it is most important that the treatment dishes are equally aligned to the wind direction, and because this continually fluctuates the dishes are usually presented on the opposite sides of a circular sheet of glass supported on a turntable that is slowly rotated (Ribbands, 1955; Butler et al., 1969; Ferguson and Free, 1979).

Soon after geraniol had been identified as a component it was shown that its attractiveness to foragers was insufficient to account for the response to Nasonov pheromone (Free, 1962; Boch and Shearer, 1964). A mixture of geraniol with geranic and nerolic acids was more effective but still failed to match the attractiveness of the Nasonov pheromone (Boch and Shearer, 1964). The citral component proved to be most attractive (Weaver *et al.*, 1964). Butler and Calam (1969) used a mixture of citral and geraniol and found that, provided the amount of citral equalled or exceeded the amount of geraniol, it was almost as attractive as Nasonov pheromone wiped on to plaster blocks; they concluded that geranic and nerolic acid were of little consequence. Shearer and Boch (1966) reported that (E)- and (Z)-citrals were equally attractive, both on their own and in combination with geraniol. However, tests showed that nerolic acid on its own was more attractive than geranic acid on its own; the addition of nerolic acid to geraniol or the citrals increased their attractiveness, but the addition of geranic acid did not. A mixture of geraniol (50 μg), citral (100 μg) and nerolic acid (200 μg) was as effective as small pieces of filter paper wiped over the Nasonov glands of 20–30 foraging bees. Preliminary tests by Free (1981) supported these results; either geraniol or (E)-citral alone were significantly attractive to searching foragers, and removing either geraniol or nerolic acid from a synthetic mixture of Nasonov components (in approximately natural proportions) significantly diminished its attraction. The presence of nerol and (E,E)-farnesol in a mixture of components sometimes diminished the collection of pollen substitute in the hive and the collection of water and sucrose syrup both in the hive and the field (Free *et al.*, 1983a). Somewhat contrary results were obtained by Williams *et al.* (1981) who found that each of the seven components was attractive, although to varying extents, and each could contribute to the attractiveness of a mixture in natural proportions. The order of attractiveness of the single components was (E)-citral, geranic acid, nerolic acid, geraniol, nerol, (Z)-citral and (E,E)-farnesol. Removal of individual components from the mixture diminished attractiveness in the following descending order of effectiveness: (E)-citral, (E,E)-farnesol, geranic acid, nerol, nerolic acid, geraniol, (Z)-citral.

It is difficult to summarize these results and clearly more experiments to resolve the differences are needed. All authors are agreed on the importance of (E)-citral and most on the importance of geraniol. Some tests suggest that nerolic acid is important and more so than geranic acid (this agrees with findings in another type of bioassay, page 123), while others suggest a reverse preference or that these acids have little or no attractiveness. These diverse results tend to emphasize the complexity of the 'language' involved and the difficulty in solving it by the means we have available.

Appraisal of bioassays

It is not easy to devise bioassays to determine the function of the components of the Nasonov pheromone when they probably interact in a complex way and perhaps have different functions in different behavioural situations. It is

Bioassay: A swarm clustered on a support in a cage containing empty hives with Nasonov lures

possible that some components are especially effective in stimulating alighting. Nearly all methods used are open to some criticism; field experiments on foraging using different odours or combinations of odours at adjacent dishes must inevitably lead to some contamination and mixing with unknown influences on the results. The 'roundabout' used by Ferguson *et al.* (1979) and Free *et al.* (1981a, 1982a) must lead to a complex mixture of components in the air in the pathway of the revolving chemicals, with quite unpredictable consequences. In the clustering and hive entrance bioassays (but not the foraging) bees responding to Nasonov pheromone are induced to release it themselves.

Even slight differences in the methods used could have significant influences on the results. It is particularly difficult to perform experiments that are free from diversions and chemical contamination, especially when the observer needs to be in close proximity as in the foraging bioassays.

However, the degree of agreement from the very different bioassays described above is fortunately often encouraging. Thus, it has been found that nerolic acid, (E)-citral and geraniol are the most important components for inducing clustering, stimulating pheromone release at the hive entrance and (usually) for attracting foragers. The presence of geranic acid has been found to diminish the attractiveness of a mixture to clustering bees and to foraging bees, although it did stimulate Nasonov gland exposure. Nerol and farnesol were not important in inducing clustering or in inducing Nasonov gland exposure at the hive entrance and in these circumstances they even tended to have a repellent effect. Farnesol, which has a relatively high molecular weight and is relatively abundant (Fig. 13.1), may well act as a keeper substance. It must always be borne in mind that in different behavioural situations bees may perhaps respond to different components or mixtures of components.

Beekeeping applications

The Nasonov pheromone lure

The results from the bioassay on cluster formation led to the development of a Nasonov pheromone lure. Each lure consists of a polyethylene vial (30 mm long, 15 mm diameter), loaded with a 1:1:1 mixture (10 mg of each) in hexane (100 μl) of (a) (E)-and (Z)-citrals, (b) geraniol and (c) nerolic and geranic acids. This mixture has the geraniol equivalent of about 5000 workers. The vial has a push-in cap which is closed immediately the chemicals have been inserted; they are absorbed into the walls and pass to the outside by diffusion.

To test the lures caged swarms were given the choice of empty hives, each

Bioassay: Bees entering hive containing Nasonov lure

Table 13.3   Percentage of hives, with and without lures, occupied by swarms.

Location	Hives without lure	Hives with Nasonov lure
Southern England (Free et al., 1981b)	0	24
New York State (Lesher and Morse, 1983)	5	32
Kenya (Kigatiira et al., 1986)	12	42

Table 13.4   Choice made by swarms for empty hives in Kenya (after Kigatiira et al., 1986).

	Hives without lure	Hives with Nasonov lure	Hives with Nasonov and 9-ODA lure
Total number of hives present	24	24	24
Number of hives occupied			
Kimbo	2	5	6
Ngong	1	2	4
Lewa	0	3	4
Athi	0	0	0
Total number of hives occupied	3	10	14

with a lure just inside the entrance, and unbaited empty hives. In 39 out of 40 trials swarms occupied a hive with a lure, and usually chose the hive most visited by scout bees. Frequently there were more than 50 scout bees at the entrance of the preferred hive immediately before the swarm arrived (Free et al., 1981b).

Decreasing the amount of chemicals in a lure decreased its effectiveness, but increasing the amount of chemicals failed to increase its attractiveness to swarms, probably because the walls of the lure were already saturated with chemicals and increasing the amount of chemicals failed to increase the amount released. However, increasing the number of lures in an empty hive increased the likelihood that it was chosen by caged swarms. Experiments in cages also indicated that the attractiveness to swarms of empty new hives, empty hives previously occupied by colonies, and empty hives that contained old empty comb were all enhanced by the presence of lures (Free et al., 1984b).

In field trials pairs of Langstroth hives, each with two empty used combs, were placed 2 m apart in 50 locations in Southern England and a lure was fixed just above the entrance inside one hive of each pair. Of the 100 hives, 12 with lures, but none without, were occupied by swarms (Free et al., 1981b). Similar tests in New York State (Lesher and Morse, 1983) and Kenya (Kigatiira et al., 1986), gave similar results (Table 13.3). Simple and inexpensive traps containing the Nasonov lure have recently proved to be an effective and competitive means of capturing honey bee swarms near Tucson, Arizona (Schmidt and Thoenes, 1987).

In tests with caged swarms the addition of one queen-equivalent of 9-ODA to the lure increased its attractiveness. However, in field tests in England the presence of this queen pheromone component had no apparent effect (Free *et al.*, 1984b). Perhaps in the unnatural environment of the cage the bees were disorientated and the presence of 9-ODA was of significance, whereas its presence was irrelevant to scout bees in the field. These results have been confirmed in Kenya (Kigatiira *et al.*, 1986), where empty hives provided with lures containing synthetic Nasonov pheromone plus 9-ODA failed to attract significantly more swarms than standard lures. (Table 13.4)

The lure is probably successful in attracting swarms to unoccupied hives because it assists in the discovery of such hives by scout bees (page 116). It may also encourage scout bees to communicate the location of the site.

Because swarming is difficult to eradicate, even by intensive beekeeping methods, Nasonov lures should be valuable in attracting swarms wherever beekeeping is practised. They should be especially useful in tropical countries where beekeepers depend upon migrating colonies to occupy their hives. At times when colonies might be expected to migrate or abscond, the presence of lures at the hive entrance might inhibit them from doing so. Similarly, beekeepers might use the lures as an aid in hiving a newly captured swarm. In Central and South America lures could be used to help 'trap' undesirable colonies of 'Africanized' bees which could then be destroyed or requeened, if necessary diminishing their aggressiveness while requeening by prolonged exposure to their own alarm pheromones (page 151–3).

Lures might also be useful in attracting swarms to cluster at particular sites (e.g. low branches from which swarms can easily be collected) soon after leaving their parent colonies. It is unlikely that they will be successful in inducing clustered swarms to enter unoccupied hives placed a few metres from them, as it seems probable (Kigatiira, 1984) that scout bees tend to favour nest sites a considerable distance away, and so reduce competition between the swarm and its parent colony for available forage. It is also possible that scout bees have located nest sites before the swarm has left its parent colony.

Increased understanding of swarming behaviour should lead to a more efficient use of the lure and elucidation of conditions under which it is likely to succeed. The chemical composition of the lure was developed especially to attract bees to join a swarm cluster. The function of all seven individual components is unknown; in different circumstances a bee may respond to different components, or to different proportions of them. Thus, a different mixture would perhaps be more effective in attracting scout bees to potential nest sites.

The Nasonov lure has uses other than attracting swarms. It can be used to trap stray honeybees in locations where they are not wanted, such as food processing factories, and glasshouses where pollination of flowering crops must be carefully controlled or prevented (Free *et al.*, 1984c). In these circumstances lures incorporating the queen pheromone component 9-ODA are especially effective, probably because many of the trapped bees have been isolated from their colonies for some hours. Bees often become disorientated, lost and die when their hives are transferred to glasshouses to pollinate the crops growing in them. Placing Nasonov lures at the hive entrances should help them return home.

Swarms typically choose their own queen rather than a foreign queen (page 105) but this discrimination is largely lost in the presence of a foreign odour, and is reversed when the foreign queen is scented with an artificial blend of Nasonov components (Boch and Morse, 1981). Therefore the Nasonov lure could be useful when introducing adult or immature queens to colonies, particularly in difficult circumstances. It might also reduce hostility when a beekeeper needs to unite two colonies.

The presence of lures had been shown (Free *et al.*, 1983a) to increase the consumption of artificial food (soya bean flour and brewers yeast mixture, pollen, skimmed milk, sugar syrup) placed immediately on top of the brood combs of a colony. They could therefore aid colony development and so improve honey production and pollination in the spring and early summer. The lures also attract bees to consume water. They could therefore be used to condition bees to collect water provided by the beekeeper near to hives, so that they would be less likely to visit water sources (e.g. taps, swimming pools or ornamental ponds) where their presence is undesirable. Attracting bees to water provided inside the hive could be especially valuable in cold weather and early in the year, when foraging can be hazardous.

Lures could be usefully developed to attract the Asian species of honeybees and especially *A. dorsata* but this must await the chemical identification of their Nasonov pheromones.

Attracting foragers to crops needing pollination

If Nasonov gland odour were applied to flowers it would probably attract 'scout' bees, in addition to the recruits directed to the flowers by the dances of successful foragers (Free, 1968a). Hence it has been suggested (Boch and Shearer, 1965a; Free, 1970a, 1978) that if Nasonov pheromone could be synthesized and applied economically, it might be used to increase the number of bees visiting agricultural crops needing pollination and so increase seed yield, especially of those crops which are marginally attractive, or less attractive than competing crops nearby.

Despite the potential importance of this, few experiments have been made and progress is slight. Butler *et al.*, (1971) sprayed a mixture of one part citral and 25 parts geraniol to plots containing about 400 dandelion (*Taraxacum officianale*) flower heads; visitation by honeybees was increased compared to unsprayed plots in two experiments (by 54% and 89%) but not in a third. Waller (1970) sprayed small plots of alfalfa (*Medicago sativa*) with geraniol and citral. On seven occasions the components were provided in dilute sugar syrup and on five occasions they were applied in water. When applied in syrup, plots sprayed with citral or geraniol alone always attracted more bees than when syrup only was applied, and on five of the seven occasions a combination of geraniol and citral attracted more bees than either component alone. In contrast when applied in water, either component on its own, or both together often failed to attract more bees than water alone. Oddly, citral seemed to exert the greater attraction in the absence of syrup, and geraniol did so in the presence of syrup. Perhaps syrup had a differential effect on the volatility of the two components.

It is difficult to explain these results. Possibly bees were unable to obtain much nectar from the alfalfa crop (see Free, 1970a; McGregor, 1976) and the added odours only became significant when food in the form of the added sugar syrup was made available. Perhaps this indicates the futility of attempting to attract bees to crops from which they can obtain little or nothing. The greater attractiveness of the two components when together (when applied in syrup but not water) probably arose because bees that became conditioned to either odour singly were also attracted to it when in combination with the other (Waller, 1973).

Williams *et al.* (1981) sprayed synthetic Nasonov pheromone, consisting of all seven components in natural proportions on to 25 cut stems of rosebay willow-herb (*Chamaenerion angustifolium*) and found that in all experiments (in which the amount of pheromone varied from 10 to 1000 gland-equivalents per plant) the sprayed stems received more visits by honeybees than 25 unsprayed control stems. No further results of applying a synthetic Nasonov mixture to crops have been reported. When such experiments are made the effect of varying the proportion of the components present, and in particular increasing the amount of E-citral and of omitting nerol and farnesol should be tested.

Application of synthetic pheromone will only be effective if the bees are in a condition to respond. Attraction to crops sprayed with Nasonov pheromone components would probably be enhanced if the colonies for pollination were taken to the crops concerned and released as spraying commences so the 'scout' bees are drawn to the synthetic pheromone on their first flights in the new location, and before they have become conditioned to forage elsewhere.

Even if it is possible to use synthetic pheromone to increase bee visitation to marginally attractive crops it may be difficult to maintain consistently the population of foragers. A slow-release formulation of synthetic pheromone could be helpful in maintaining a continuous attraction.

Sustained foraging is dependent upon adequate reward in nectar and pollen, and unless this is forthcoming the initial attraction by Nasonov pheromone may well be only transitory.

Chapter fourteen

ALARM AND AGGRESSION
PHEROMONES

Introduction

The brood and stores of honey and pollen inside a honeybee nest offer a tempting prize to many animals including man and foraging honeybees from other colonies. An effective defence is vital. Because the word 'defence' has a passive implication, and indeed some defensive behaviour (e.g. attempts to escape or otherwise avoid conflict) is distinctly non-combative, honeybees attacking in defence of their colonies (counter-offensive attacks) and in other circumstances will be referred to as 'aggressive'.

An alert guard bee (*Apis mellifera*)

Guard bees are usually present at the nest entrance. These are bees that have finished house duty and have not yet undertaken foraging; they probably have a low stinging threshold. But guards are relatively few compared to the colony population, and when danger is imminent alarm pheromones serve the dual functions of being used to muster help and to direct the attack. This can result in synchronized attacks often by more than 100 bees against an intruder.

This has long been known. Huber (1814) found that when he placed an excised sting at the entrance to a hive the bees became agitated and attacked him and the sting. He obtained a similar reaction when he presented the odour alone of aggressive bees that had been protruding their stings. Even earlier Butler (1609) had described how a single bee sting in the clothing or skin can attract many other bees to attack.

Alerting behaviour

When a guard bee is disturbed, such as by tapping her briefly with a finger, she will often raise her abdomen and release alarm pheromone by opening her sting chamber and protruding her sting (see photograph overleaf). While beating her wings and so aiding dispersal of the pheromone, she will run into the hive and alert her colony. After a few seconds many excited bees may rush out of the hive entrance and run about jerkily in circles or zigzags, or stop and assume a characteristic tense and aggressive posture with a slightly raised body, wings extended, mandibles agape and antennae waving. They are ready to fly to the attack at the slightest further provocation. These two effects, alerting and activation, are characteristic of alarm pheromones. The number of bees at the entrance increases as additional bees are alerted and recruited. (Maschwitz, 1964).

The alarm pheromone of the sting chamber (Fig. 2.3) accumulates beneath the sting shaft setose membrane (the membranous venter of the 9th segment; Snodgrass, 1956) which folds over the bulb of the sting shaft (Gunnison and Morse, 1968). According to Ghent and Gary (1962), the secretion seems to be produced by two masses of glandular cells lying against the inner surfaces of the quadrate plates and the secretion accumulates in the reservoir between the sting shaft membrane and the bulbous base of the sting shaft; it does not appear to be produced by any of the four known glands (the venom gland, the Dufour gland, paired Koschevnikov glands or paired sting sheath glands) associated with the sting (Maschwitz, 1964).

When a honeybee extrudes her sting shaft it is wiped by the hairy under surface of the sting shaft membrane, so the sting shaft may be coated with alarm pheromone, and the sting shaft membrane itself is protruded as a hood over the base of the sting, so facilitating dispersal of the remaining material. It is not known how quickly alarm pheromone is dispersed, how soon it is

A worker honeybee (*Apis mellifera*) releasing alarm pheromone at the entrance to its hive. (Photograph: U. Maschwitz).

replenished, or indeed whether a worker continues to assume the alarm releasing posture when its alarm pheromone supply is exhausted.

Huber (1814) observed that when he agitated bees that were in a semi-torpid condition they protruded their stings upon which drops of venom appeared. Bees of a winter cluster frequently extrude their stings when disturbed (Ghent and Gary, 1962) presumably releasing alarm pheromone, and in sub-zero temperatures when the bees on the cluster periphery are too cold to fly their behaviour alerts warmer bees in the cluster interior to fly to the attack (Morse, 1966).

Much still needs to be discovered about the conditions under which bees release alarm pheromone, and whether it is usually only when large intruders (mammals or birds) constitute a danger.

Releasing stinging

Alerted workers need to search and discover the enemy before their aggression can be released. Furthermore, although the odour of alarm pheromone has alerted the bees and made them ready to sting they need an additional stimulus to do so. Certain characteristics of the enemy, especially the odour, jerky movement, hairy body covering enable bees to recognize it and provoke attack (Free, 1961b). Presumably these characteristics are sometimes great enough to provoke attack without the bees first being alerted by alarm pheromones. Pheromone from the sting gland is also used to mark an enemy and make it a more obvious target. A bee sting is normally retracted within its sting chamber but when attacking it is protruded ready to thrust into an enemy. The shaft of a sting is barbed and a bee is unable to withdraw it from the skin of vertebrates, so the sting, together with associated motor apparatus and glands are severed from the bee as it attempts to fly away and are left attached to the enemy. The severed sting apparatus continues to pump venom into the victim and alarm pheromone is dispersed from the exposed under-surface of the sting shaft membrane (Ghent and Gary, 1962). Venom itself does not stimulate stinging (Free and Simpson, 1968) but under some circumstances it may contain alarm pheromone (Gunnison, 1966), the conditions under which this occurs are not clear. Probably the alarm pheromone is dispersed from a severed sting within a few minutes. We do not know the relative amount of alarm pheromone released when a bee is alerting others and when its sting is embedded in an adversary. Perhaps the composition of alarm pheromone differs in these two circumstances.

It is not clear to what extent the workers recruited by alarm pheromones subsequently release it themselves, nor to what extent bees that release alarm pheromone subsequently fly to the attack, perhaps activated by their own pheromone.

Marking an intruder with alarm pheromone has the biological advantage of directing the attacks of other defending bees toward it, so once an enemy or object has been stung, its chances of receiving additional stings is greatly increased (Free, 1961b; Ghent and Gary, 1962). Colony odour may also assist bees to orientate to the target as bees preferred to attack cotton balls that had been given the opportunity to acquire the odour of their own colony rather than the odour of a foreign colony (Free, 1961b).

A sting left behind in flesh

Table 14.1 Summary of known functions of honeybee alarm pheromone components.

Source	Alarm pheromone component	Releases alerting in cages[1]	Releases alerting at hive entrance[2]	Repels at hive entrance[2]	Releases stinging[2]	Repels clustering bees[2,3]	Inhibits scenting[2,3]	Inhibits foraging activity[2]	Encourages foraging activity[2]	Repels foragers[3]	Attracts foragers[4]
Mandibular glands	2-Heptanone[a]	*			+	*	*	+		+	
Sting chamber	Isopentyl acetate[b]	*	*		+	*	*	*		+	
	n-Butyl acetate[c]	+	+		+ *			*		+	
	n-Hexyl acetate[c]	+	+		*					+	
	n-Octyl acetate[c]	+		+		*	*	+ *		*	
	2-Nonyl acetate[d]		+			*	*	+		+	
	n-Decyl acetate[c]										
	Benzyl acetate[c]	+	+	+						*	
	Eicosanol acetate[g]		+								
	Octanoic acid[g]		+			+	+	*		+	
	Benzoic acid[g]										
	3,3-Dimethyl acrylic acid[g]										
	1-Butanol[d]		+		+	+	+ *	+ *			
	1-Pentanol[d]		+	+	+ *	+	*	+ *	*	+	
	Isopentyl alcohol[c]	+						+ *		+	
	2-Heptanol[d]		+							+	
	1-Octanol[d]		+		+	+	+	+	+	+	
	2-Nonanol[c]	*			+				+ +	+ *	
	Octadecan-1-ol[g]			+	+						
	9-Octadecen-1-ol[g]				+	*	*				*
	Phenol			+ *	+				+	*	
	p-Cresol[g]			+		*					
	Benzyl alcohol[c]					+				+	
	Eicosan-1-ol[g]										
	(Z)-11-Eicosen-1-ol[d,e]										+

* : strongly,
+ : significantly.

Identified by:
a Shearer and Boch, 1965;
b Boch et al., 1962
c Blum et al., 1978
d Blum, 1982;
e Pickett et al., 1982;
f Collins and Blum, 1983;
g Blum, 1984;

From:
1 Collins and Blum, 1982;
2 Free et al., 1987e;
3 Free et al., 1983a;
4 Free et al., 1982b;

However, even though coated with alarm pheromone an intruder or object must move or be moved, preferably jerkily (Free, 1961b; Ghent and Gary, 1962) before it releases stinging. The need for this supplementary stimulus prevents wastage of bees' lives in attacking a foe that has already been killed.

Identification and characteristics of sting gland components

Boch et al. (1962) identified a main component of the alarm pheromone of the sting gland as isopentyl acetate. It is absent in queens and young worker bees up to three days old, and there is little in bees up to a week old. The amount present then rapidly increases until it reaches a maximum of 4–5 μg per bee when 2–3 weeks old. The amount then decreases to about 2 μg per sting as bees become established foragers. The amount produced depends more on the state of physiological development of the bees than on its chronological age. The maximum amount is present when bees are usually undertaking guard duty and beginning to forage. Bees that are confined in cages since emergence produce none or very little (Boch and Shearer, 1966).

Tests with isopentyl acetate showed that it both alerted bees and induced them to sting but in neither function was it as effective as the complete alarm pheromone (Boch et al., 1962, 1970; Free and Simpson, 1968).

Several more components, which are primarily acetates and alcohols of low molecular weight, have been identified (Blum et al., 1978; Blum, 1982, 1984; Pickett et al., 1982) and attempts have been made to discover their functions. Behavioural responses to several of the identified alarm pheromone components are summarized in Table 14.1. The functions of isopentyl butyrate, 2-heptyl acetate, 1-acetoxy-2-octene, 1-acetoxy-2-nonene, 2-undecanol (Blum, 1982), 1-acetoxy-2-decene, 2-nonadecen-1-ol (Blum, 1984) have yet to be determined.

Most of the active components in common with those of alarm pheromones of other insects, are characteristically rather simple molecules, with low molecular weights ranging between 100 to 200, and are very volatile. Hence they can broadcast a message very rapidly and elicit a fast response critical for colony survival. But they are unstable and do not persist for long in the region where they have been emitted unless the danger stimulus is applied continuously.

Bioassays

Because most of the components dissipate rapidly it is difficult to control the amount tested. Furthermore, bioassays of alarm pheromones are difficult because reacting bees themselves release alarm pheromone. Both the release of alarm pheromones and the response to them will vary with the weather, foraging conditions and any recent disturbance. The main bioassays used and the significant results obtained are given below.

Alerting at the hive entrance bioassay

The compound is placed on a stationary lure (e.g. a small cork, cotton ball or piece of filter paper) on the alighting board at the entrance to a hive and the number of bees alerted by it and the behaviour of bees toward it compared with the effect of an untreated control (e.g. Boch *et al.*, 1962, 1970; Boch and Shearer, 1965b; Free *et al.*, 1987e).

Several components (isopentyl acetate, n-butyl acetate, hexyl acetate, 2-nonyl acetate, eicosanol acetate, octanoic acid, 1-butanol, 1-pentanol, 2-heptanol, 1-octanol) alert the colony and attract alerted bees, while others (e.g. octyl acetate, *p*-cresol and benzol alcohol) fail to alert the colony and repel bees.

The reaction of bees to the alarm pheromone components sometimes increases with concentration and may sometimes actually differ with increasing concentration. For example, when isopentyl acetate is presented at the hive entrance (Boch and Shearer, 1971) there is, at first, with increase in concentration a marked increase in the number of responding bees. But, at excessive concentration the bees hesitate in their approach to the odour source, they stop at some distance from it, and make short probing movements forward and backward, turn away and run about in an erratic manner. 1-Pentanol which is attractive at low concentrations also becomes repellent at high concentrations (Free *et al.*, 1987e).

Alerting in cages bioassay

Small groups (10–50) of queenless bees in cages are exposed to the compound and the number that respond by increased locomotion and partial extension of the wings, and the intensity of the response, is recorded.

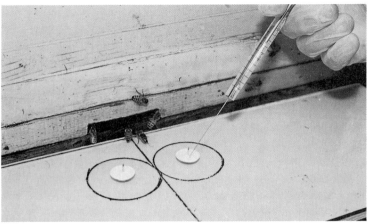

Bioassay: Alarm pheromone component being applied to filter paper disc to determine alerting response

Bees of four weeks old react faster than newly-emerged or six-week old bees but the duration of response is similar. Young bees are preferred because they do not confuse the results by producing their own alarm pheromone (Collins and Rothenbuhler, 1978; Collins, 1980).

The components most effective in eliciting reactions from caged bees are isopentyl acetate and 2-nonanol. Two components tested, 1-decanol and phenol, fail to produce a reaction (Collins, 1981; Collins and Blum, 1982, 1983).

Stinging bioassay

Pairs of balls of cotton or leather, or squares of leather, one of which is treated with the compound being tested, are jerked in front of a hive entrance. Records are made of the first to be stung, and the number of stings in each (e.g. Free, 1961b; Michener, 1972; Stort, 1974). Attempts have been made to devise a more standardized test (Collins and Kubasek, 1982) in which a hive is first given a physical jolt of constant force to arouse bees to the level at which they are likely to sting, followed by provision of a regularly moving target of suede leather to stimulate stinging.

Many components stimulate attack and release stinging. n-Butyl acetate and 1-pentanol appear to be especially effective (Free et al., 1983b; 1987e; Al-Sa'ad et al., 1987).

Nasonov gland exposure and clustering bioassays

Small groups of bees, confined in cages, are subjected to alarm pheromone components. Many of the components (e.g. 2-heptanol, 1-pentanol, n-hexyl

Bioassay: Worker bees attracted and alerted by alarm pheromone component applied to filter paper disc

acetate, benzyl acetate and 2-nonanol) induce the bees to expose their Nasonov glands and fan their wings (Collins, 1980, 1981; Collins and Blum, 1982). The significance of this behaviour is obscure but such queenless, homeless bees readily engage in scenting behaviour in response to disturbance. Chemical compounds, not known in honeybee pheromones, may give a similar response (Collins and Blum, 1983), suggesting that it is related to the presence of a strange odour rather than being a specific response to a pheromone component.

In more natural circumstances alarm pheromone has the opposite effect. When bees of a swarm settle around their queen they expose their Nasonov glands and fan their wings. Releasing isopentyl acetate near the queen inhibits scenting behaviour and no additional workers join her (Morse, 1972). When a swarm is given a strange queen a few of the workers soon clasp her with their mandibles, protrude their stings and probably release alarm pheromone; release of Nasonov pheromone and attraction of additional workers ceases (Boch and Morse, 1974).

Tests of several alarm pheromone components on bees that are fanning and scenting at their hive entrance, as they move into the hive to join the cluster, showed that some inhibited scenting (1-pentanol, (Z)-11-eicosen-1-ol) some inhibited clustering (phenol, p-cresol) some inhibited both (isopentyl acetate, n-octyl acetate, 2-nonyl acetate, octanoic acid, 1-butanol, 1-octanol), and others had no effect (Free et $al.$, 1983b, 1987e).

Foraging activity bioassay

Simultaneous counts are made of foragers entering and leaving a pair of hives. The odour of an alarm pheromone component is dispensed just outside the entrance of one hive but not the other, and the counts are continued (Free et $al.$, 1987e).

Most components discourage foraging activity, especially isopentyl acetate, n-butyl acetate, n-octyl acetate, octanoic acid, isopentyl alcohol and 1–octanol; as a result the defence force available is increased. In contrast, adapting bees to pheromone dispensed within their hive (page 151) does not influence their foraging (Al-Sa'ad et $al.$, 1985). A few components, especially 1-butanol, appear to encourage foraging.

Foraging bioassay

Foragers are trained to collect sucrose syrup from a dish placed in the centre of a circular table. The dish is removed and replaced by two without syrup positioned on opposite sides of the table. One dish contains the component under test; the other dish serves as a control. The would-be foragers landing at each are counted (e.g. Ferguson and Free, 1979; Free et $al.$, 1987e).

Many of the components repel foragers, but especially n-octyl acetate, benzyl acetate, 2-nonanol and phenol. (Free *et al.*, 1987e). Two components, (Z)-11-eicosen-1-ol (Free *et al.*, 1982b) and 9-octadecen-1-ol (Free *et al.*, 1987e), are attractive to foragers.

Multi-functions of sting gland components

The situation is very complex. It is evident that at least some components are specialized for different functions or are more successful and useful in one context than another. For example, only one component, n-octyl acetate, appears to be consistently inhibitory or repellent in all the behavioural contexts in which it has been tested. Even when two components share a function their relative effectiveness may be very different. Furthermore, the relative amounts and effectiveness of the different alarm components and so the quantity and quality of information conveyed probably changes with the age and occupation of the bees. Boch and Shearer (1966) showed that five components (unidentified) of the sting gland varied in amount with the age of the bee in a similar way to isopentyl acetate (page 139), whereas six other components did not occur in bees until they were two weeks old, and thereafter increased in amount as the bees grew older. Therefore the alarm pheromone of bees of different ages and specialized in different colony occupations may convey subtly or even markedly different messages.

Because the components have different volatilities the message may also change with the duration of exposure of the alarm pheromone and the distance from the bee concerned, so giving a spatially patterned message. However, perhaps some of the less volatile oily components, and especially the alkanes and alkenes (Blum *et al.*, 1978) help to retain some of the more volatile. (Z)-11-eicosen-1-ol may sometimes serve such a function (Blum, 1982; Pickett *et al.*, 1982). Components of low volatility may be especially useful in this context by providing a longer-lasting signal when alarm pheromone is used to mark the foe rather than for alerting only.

Tests so far have concentrated on testing the alarm components singly. Although the results provide valuable indicators of function they should be treated with caution. The complete roles of the different components may not become apparent until they are tested in various combinations and in various proportions.

Mandibular gland pheromone

Stinging bees often grip an enemy with their mandibles and Free (1961b) suggested that while doing so they might deposit an alerting substance. Maschwitz (1964) found that at a hive entrance more bees examined the mandibular glands (Fig. 2.3) of worker honeybees squashed on filter paper than examined unscented filter paper, but still more examined filter paper impregnated with substance from the sting apparatus. Consequently, he suggested that the mandibular glands produce an alerting pheromone, although a less effective one than the sting pheromone. Shearer and Boch (1965) identified 2-heptanone from the mandibular gland secretion, and when filter papers or small corks carrying 2-heptanone were put at the hive entrance, the guard bees were alerted and attacked them. However, in a later series of tests Boch et al., (1970) found that isopentyl acetate was 20–69 times more potent than 2-heptanone in alerting bees, so such a function for 2-heptanone seems doubtful.

In contrast, Free and Simpson (1968) found that 2-heptanone was as effective as isopentyl acetate in eliciting attacks on cotton balls. As the mandibles are used for grasping an intruder it seems likely that the main function of 2-heptanone is to label the intruder to be attacked.

The balance between isopentyl acetate and 2-heptanone changes during the life of the bee; when bees begin foraging the amount of isopentyl acetate diminishes while the amount of 2-heptanone continues to increase (Boch et al., 1970).

Worker bees' mandibular glands secrete 10-hydroxy-2-decenoic acid, the main lipoid content of larval food, as well as 2-heptanone. The discovery by Boch and Shearer (1967) that no, or very little, 2-heptanone is produced in the mandibular glands of worker bees undertaking nest duties refuted earlier suggestions that 2-heptanone is secreted in newly discharged larval food to deter nurse bees giving more food to larvae that had just been fed. The production of 2-heptanone is linked to the activities of the worker bee; it appears in the mandibular glands of bees when they become guards and foragers, between two and three weeks old, and then rapidly increases with age to a maximum of about 40 μg per bee (Boch and Shearer, 1967; Crewe and Hastings, 1976). Bees confined to cages and unable to forage and guard show no such increase in 2-heptanone production.

Other functions for 2-heptanone have been suggested, but not demonstrated, that would conform with its major production by older bees. Crushed bees' heads (Simpson, 1966) and 2-heptanone (Butler, 1966b; Ferguson and Free 1979) are repellent to bees that are foraging on sugar syrup. However, the repellent effect of 2-heptanone on foraging bees may merely reflect conditioned foragers' reactions to a physiological stimulus that is being presented outside its normal context. Simpson (1966) suggested that under natural conditions it might deter would-be robber bees, which are merely bees foraging for honey from strange honeybee colonies. Alert guard bees at the nest entrance in a threatening posture can be seen to open and close their mandibles and could be releasing pheromone. Insects, other than honeybees could conceivably be repelled by 2-heptanone.

Bees that have visited unrewarding flowers mark them to discourage further visits (page 112; Núñez, 1967; Free and Williams, 1983); the pheromone concerned has not yet been identified; perhaps it is 2-heptanone.

The volatiles of previously used comb encourage bees to store food in it (Free and Williams, 1972; Rinderer, 1981). Strangely, increased hoarding also occurs in comb that has been exposed to 2-heptanone (Rinderer, 1982); no satisfactory explanation has been put forward to account for this, especially as the 2-heptanone would be expected to have dissipated soon after the experiments began. When small groups of bees, kept in cages, are exposed to 2-heptanone some of them expose their Nasonov glands (Collins, 1980, 1981); perhaps it is the Nasonov pheromone released in these circumstances that stimulates hoarding.

When a beekeeper introduces a strange queen to a colony she is sometimes suddenly attacked by many workers which encompass her in a ball. This attack may be initiated directly or indirectly by release of 2-heptanone from workers that are examining her (Yadava & Smith, 1971). When worker bees are inspecting a foreign queen they often bite her legs and wings; they may be marking her with 2-heptanone. When 2-heptanone is placed in the vicinity of a queen at which bees of a swarm are gathering they soon cease to expose their Nasonov glands and no new bees are attracted to her (Morse, 1972). A suggestion, (Yadava and Smith, 1971) that in response to 2-heptanone the queen produces a 'stress' pheromone that induces bees to attack her has not been confirmed (Ambrose, 1975).

However, despite the conjectures that have been made it is doubtful whether 2-heptanone serves a major purpose other than guiding defending bees to a target. 2-heptanone is about as effective at releasing stinging as the entire mandibular gland contents, and there is no evidence that any other substance from the mandibular glands is involved in colony defence.

Effect of empty comb on defence

The presence of empty comb enhances the level of defensive behaviour. Colonies housed in hives with large areas of empty comb respond to targets twice as fast and sting the targets twice as much as colonies with little empty comb (Collins and Rinderer, 1985). The presence of unidentified pheromones in empty comb, which also encourage hoarding and foraging (page 81), must therefore make the bees physiologically or psychologically more responsive to alarm.

As Collins and Rinderer (1985) pointed out, a prime requirement of established feral colonies with much empty comb is to defend such small stores of honey as they do possess. Factors lowering the threshold of response to intruders are especially advantageous in such situations.

Aggressivenes of different races of *Apis mellifera*

It is well known that colonies differ greatly in the eagerness with which they defend themselves against intruders. These differences may be linked to the amount of alarm pheromones their bees produce or release, or to the threshold of response of their bees to alarm pheromones.

For example, Free and Williams (1979) were unable to induce the docile *A. florea* colonies they were studying in Oman to respond to isopentyl acetate or to 2-heptanone although *A. florea* colonies in the Philippines readily did so (Morse *et al.*, 1967).

Chromatograms of sting extracts of *Apis mellifera* from the Philippines, Italy and N. America showed exactly the same composition but bees of an Italian race (var. *ligustica*) contained more isopentyl acetate per sting than bees of a Caucasian race (var. *caucasica*) (Morse *et al.*, 1967).

Collins and Rothenbuhler (1978) found that in a laboratory test bees from different inbred lines differed in their speed of reaction to isopentyl acetate and in their intensity of response. However, the relative response of different colonies to different components can vary. The speed of response (seconds) of three colonies (A, B and C) to three different alarm pheromone components was:

	Colony		
	A	B	C
Isopentyl acetate	5.1	4.2	7.5
Isopentyl alcohol	5.1	7.6	4.9
Benzyl acetate	5.6	5.6	11.2

A. mellifera races introduced into Brazil from Africa become aggressive and sting more readily than those from Europe (Kerr *et al.*, 1974; Stort, 1974). When artificial pheromone is released at the entrance to an Africanized colony many of the bees immediately become airborne; in contrast the reaction of European bees is to remain near their hive entrance (Collins *et al.*, 1982).

These differences in behaviour of European and Africanized bees also seems to depend on the lower threshold of response of the Africanized bees. Agressiveness is not always, or closely, correlated with the amount of isopentyl acetate present (Boch and Rothenbuhler, 1974; Kerr *et al.*, 1974; Crewe and Hastings, 1976), but could perhaps be associated with the amounts or proportions of other pheromone components. Alarm pheromone in the sting chamber is not stored but is produced as needed; perhaps the rate at which it is synthesized increases with aggressiveness. Kerr *et al.*, (1974) did find a positive correlation between aggressiveness and the amount of 2-heptanone stored in the mandibular glands.

As with European races, the production of alarm pheromones in Africanized bees varies with age. Until they are about three weeks old only small amounts of 2-heptanone are produced in Africanized workers. Amounts of isopentyl acetate also increase with age, reach a peak when workers are about three weeks old and then diminish somewhat (Crewe and Hastings, 1976).

Despite their lower threshold of response to alarm pheromones, Africanized bees have a smaller number of sensilla placodeae on their antennae than Italian bees (Stort, 1974).

Other *Apis* species

Like *A. mellifera* some strains of the Indian honeybee *A. cerana* attack an intruder more readily than others. The dwarf honeybee (*A. florea*) is often quite docile (Free and Williams 1979) whereas the giant honeybee (*A. dorsata*) is often extremely aggressive and 1000–5000 bees may leave a nest within a few seconds to attack an enemy (Morse and Laigo, 1969).

Extracts of stings of all four *Apis* species contain isopentyl acetate. Samples of *A. mellifera* workers contained between 0.8 and 2.8 μg isopentyl acetate per bee, *A. cerana* workers between 0.2 μg and 1.5 μg per bee, and *A. florea* workers between 0.2 μg and 1.1 μg per bee. In contrast, four samples of *A. dorsata* foragers or guard bees contained 23.8, 58.0, 21.8 and 18.0 μg per bee respectively; even young *A. dorsata* bees contained 5.0 μg (Morse *et al.*, 1967; Koeniger *et al.*, 1979). All four honeybee species contain n-octyl acetate in addition to isopentyl acetate (Veith *et al.*, 1978).

Morse *et al.*, (1967) reported that at least three fractions other than isopentyl acetate occurred in *A. dorsata* sting extracts and were not present in other species. Veith *et al.*, (1978) and Koeniger *et al* (1979b) found that sting extracts of *A. dorsata* and *A. florea*, but of neither of the other two species, contained 2-decen-1-yl-acetate. *A. florea* and *A. dorsata* alone responded to sting extracts containing 2-decen-1-yl-acetate but no isopentyl acetate. 2-Decen-1-yl-acetate has a low volatility and so is able to alert the colony and perhaps to mark the enemy for longer periods than the highly volatile isopentyl acetate. Its presence could explain why the duration of response of colonies to extracts of stings of their own workers is two to five times as long for *A. florea* and *A. dorsata* as for *A. mellifera* and *A. cerana*.

Alarm pheromone released and dispensed by fanning bees at the entrances of *A. cerana* and *A. mellifera* colonies is able to rapidly alert bees in the interior of their nests. However, the exposed colonies of *A. dorsata* and *A. florea* are protected by a curtain of bees, several layers thick, so it would take longer for alarm pheromones to reach the active bees beneath the curtain.

A. mellifera and *A. cerana* colonies nest in sheltered locations with restricted entrances. *A. dorsata* colonies are often exposed on high branches. *A. florea* colonies are sometimes in caves or other shelters but are usually on shaded branches near the centres of bushes and trees. The large amounts of isopentyl acetate produced by *A. dorsata* and *A. florea* may be partly a

Apis dorsata colonies sharing the same tree (Bangladesh)

consequence of the exposed sites of their nests, and the speed at which alarm pheromone is dispersed.

A. dorsata colonies tend to be gregarious, several often nesting on the high branches of the same tree. Presumably the main advantage of the gregarious nesting is in providing mutual aid in defence against large predators. For this to be possible the alarm pheromones released by a threatened colony must be sufficiently abundant and durable to alert colonies some metres away.

Normally an alarm pheromone of low volatility would have the disadvantage of maintaining individuals in an alert state and distracting them from other activities when danger no longer threatens. Perhaps in the exposed situations in which *A. dorsata* and *A. florea* workers release alarm pheromone, even components of low volatility soon cease to be effective.

2-Heptanone is not found in *A. dorsata*, *A. florea* or *A. cerana* although when presented at the nests of the latter two species it elicits alarm behaviour (Morse *et al.*, 1967).

Beekeeping applications

Bee breeding

Alarm pheromones can be used in standardized tests to compare the aggressiveness of different colonies, and so can be used as a tool in breeding a more docile strain.

Repelling bees from crops

Under certain conditions as when foraging at dishes of sucrose syrup (page 142), honeybees are repelled by alarm pheromone. Attempts have been made to use synthetic alarm pheromone to repel foragers from crops (Free *et al.*, 1985b) in the hope that, if successful, it could be applied just before insecticide on those crops where bees are at risk.

In preliminary experiments isopentyl acetate and 2-heptanone were applied to alternate sunflower heads that had been arranged in a large circle. Immediately after application foraging on the treated heads was greatly diminished, but within 12 minutes had approached the pretreatment level (Fig. 14.1). These initial experiments were followed by field tests. During the first 30 minute period, after high concentrations of isopentyl acetate and 2–heptanone had been applied to field plots of oil seed rape and field bean, foraging was reduced by more than 80% and 40% respectively; but by the second 30 minute period after application most of this repellancy had disappeared (Fig. 14.2).

The transient nature of the repellancy probably arises from the high volatility of the alarm pheromones and more work is necessary to produce a mixture of greater persistence. Preparation of less volatile and more stable propheromones that would release pheromone when exposed to sunlight in

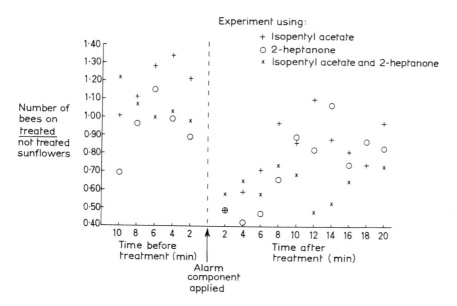

Figure 14.1 Effect of applying alarm pheromone components to sunflower heads on the proportion of honeybees visiting them compared with untreated heads (from Free *et al.*, 1985b).

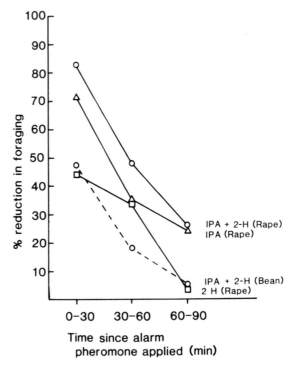

Figure 14.2 Reduction in foraging following application of high concentrations of the alarm pheromone components isopentyl acetate (IPA) and 2-heptanone (2-H) to crops of oil seed rape and field bean (summarized from Free *et al.*, 1985b).

the field has been suggested (Pickett *et al.*, 1985). A mixture of two of the more recently discovered alarm pheromone components (n-octyl acetate and benzyl acetate, Blum *et al.*, 1978) shows promising results of greater persistency when applied to sunflower heads and appears to be more repellent than a mixture of isopentyl acetate and 2-heptanone (Free *et al.*, 1987e). It has yet to be tested on field crops.

Hopefully, application of synthetic alarm pheromone a few minutes before an insecticide might help reduce any bee losses to acceptable levels. Probably it would only be necessary to apply the alarm pheromone to the borders of the field as foragers leaving or returning to the field would then become aware of it.

Protecting colonies from alien honeybees

Because alarm pheromone repels foragers, and is probably released by a colony defending its honey stores from would-be foragers of other colonies it might also help to repel such alien bees. This has yet to be shown. But if true, application of synthetic pheromone at the hive entrance could repel robbers and ensure the defenders are alert. Possibly the defending colony would in time become adapted to the alarm pheromone (page 153) and so any initial diminution of its foraging would not be prolonged. However, even if this proves to be wrong, it should be possible to repel the alien bees with those pheromone components (i.e. benzyl acetate, 2-nonanol, phenol, benzyl alcohol) that repel foragers but do not inhibit the colony's own foraging activity.

Reducing aggression

Alarm pheromones are highly volatile, and when danger is no longer present colony activities soon return to normal. So under natural conditions there is little opportunity for bees to become adapted to their own alarm pheromones. However, it has recently been demonstrated that honeybee colonies can rapidly become adapted to synthetic alarm components dispensed within their hives and as a result are less inclined to sting in defence (Al-Sa'ad et al., 1985).

The components concerned were dispensed from the wick of a small spirit lamp or other small container placed on the floor of the hive. During tests

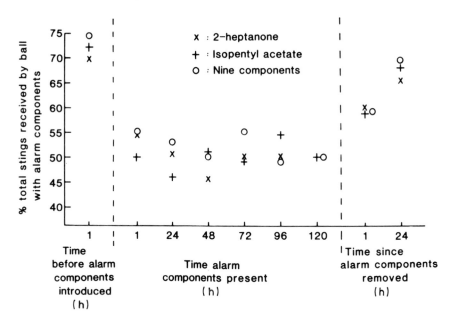

Figure 14.3 Effect of exposing colonies to alarm pheromone components (either isopentyl acetate, 2-heptanone or a mixture of 9 components) on their tendency to sting cotton wool balls treated with the components concerned compared to an untreated control ball (after Al-Sa'ad et al., 1985).

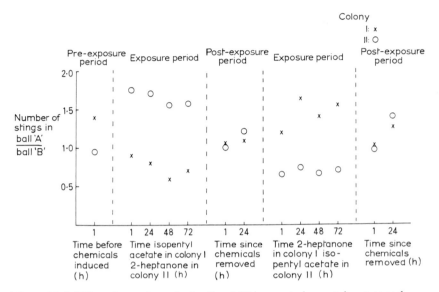

Figure 14.4 Effect of exposing colonies (I and II) in turn to isopentyl acetate and 2-heptanone on their tendency to sting cotton wool balls treated with isopentyl acetate (ball 'A') and 2-heptanone (ball 'B') (from Al-Sa'ad et al., 1985).

conducted before colonies were exposed to the alarm pheromone components, they stung cotton balls treated with alarm pheromone about twice as much as untreated balls. However, within an hour of dispensing synthetic alarm pheromone defending bees ceased to discriminate between treated and control balls, and this lack of discrimination persisted while the treatment was continued. Discrimination was only partial one hour after the synthetic pheromone dispenser had been removed, but returned to about pretreatment level after 24 hours. A mixture of components received a similar response to single components (Fig. 14.3). The results of alternately treating colonies with isopentyl acetate and 2-heptanone on their stinging responses are given in Fig. 14.4.

Colonies adapted to a mixture of components known to be most effective in releasing stinging also have a greatly reduced tendency to sting untreated targets (Experiments 1–5, Table 14.2). Incorporation of recently discovered components that release stinging (pages 138 and 141) into the mixture dispensed within the hive (Experiments 6 and 7) failed to increase its effectiveness (Free et al., 1987f).

Because bees adapted to their own alarm pheromones do not especially select targets marked with them and are less ready to sting in defence of their colonies, it should be possible to reduce the aggressiveness of colonies with a low threshold of stinging and make it easier for the beekeeper to replace their

Table 14.2 Effect of adapting honeybee colonies to synthetic alarm pheromone components on their stinging response (from: Al-Sa'ad et al., 1985; Free et al., 1987f).

	Components dispensed in colonies		*% change in stinging by group A relative to group B after:*		
Experiment	Group A	Group B	24 h	48 h	72 h
1	Isopentyl acetate and 2-heptanone	None		−34	
2	Isopentyl acetate and 2-heptanone	None		−33	
3	Isopentyl acetate, 2-heptanone & n-butyl acetate	None		−48	
4	Isopentyl acetate, 2-heptanone & n-butyl acetate	None	−36	−37	
5	Isopentyl acetate, 2-heptanone & n-butyl acetate	None	−21	−76	
6	Isopentyl acetate, 2-heptanone & n-butyl acetate	Six*	−17		
7	Isopentyl acetate, 2-heptanone & n-butyl acetate	Eight†	+6		−72

* 2-heptanone, isopentyl acetate, n-butyl acetate, 1-butanol, 1-pentanol, 2-nonanol

† 2-heptanone, isopentyl acetate, n-butyl acetate, 1-butanol, 1-pentanol, 2-nonanol, 9-octadecen-1-ol and p-cresol

queens with more docile ones. This technique may be especially useful in diminishing the aggressiveness of 'Africanized' colonies in South and Central America. However, it is unlikely that stinging will be eliminated completely by this technique because the characteristics of the intruder, especially odour and movement, themselves release stinging.

Introducing synthetic alarm pheromone into a hive without arousing undue aggression could be difficult, but perhaps this could be achieved by inserting the container into the hive entrance after dark, and coating with sugar candy or honey that part of the container on which the alarm pheromone is to be exposed; the bees themselves would soon remove this coating.

Fortunately adaptation to alarm pheromones does not appear to influence foraging activity and so would not diminish the pollinating value of honeybee colonies foraging on agricultural crops. Furthermore, it does not make guard bees less diligent at inspecting bees from their own and strange colonies at the hive entrance, so a colony's alertness is maintained and it is unlikely to be overwhelmed by intruders (Al-Sa'ad et al., 1985).

Effect of smoke on colonies

It is well known that blowing smoke into a colony's hive or nest helps to diminish aggression. This is partly because it induces many of the worker bees to ingest food from the storage cells which reduces the likelihood that they will sting (Newton 1968, 1971; Free, 1968b). However, only about half the bees engorge when their colony is smoked and aggression begins to diminish immediately smoke is applied to a colony, so it seems likely that other effects of smoke on bees are also important to the beekeeper. Smoke may also inhibit aggression by deterring bees from leaving their comb, by distracting their attention from an intruder, or by masking an intruder's alien scent. Moreover, particular components of smoke may mask the odour of alarm pheromones or prevent the antennal sensilla sensitive to alarm pheromones from functioning normally. If either of these two latter alternatives proves to be true it could lead to use of a more convenient material for pacifying colonies.

Stingless bees

Some species of Trigona (e.g. T. pectoralis and T. cupira) produce alarm pheromones that release mass attack. Different components of their alarm pheromone may have different functions. The alarm pheromone of T. pectoralis contains 2-heptanol, 2-nonanol, 2-heptanone and 2-nonanone. When exposed to 2-heptanone alone bees aggregate at the mouth of the entrance tunnel and remain motionless whereas exposure to 2-nonanone alone releases attack.

Major components of the mandibular gland secretion have been identified as nerol and geraniol for *T. subterranea* (Blum *et al.*, 1970) and 2-heptanol for *T. spinipes* (Blum, 1974).

The mandibular gland secretion of *T. gribodoi* foragers contains 8 components: 2-heptanone, nonanal, 2-nonanol, 1-nonanol, (E)- and (Z)-citral, tetradecane, and hexadecane. The glandular secretion of nurse bees contains only 4 of the components (nonanal, 1-nonanol, (E)-and (Z)-citral) and they are at much lower concentrations than in the foragers (Keeping, *et al.* 1982).

T. gribodoi nurse bees appear oblivious to mandibular gland extract but foragers respond immediately by frequent and vigorous wing-fanning and rapid, jerky walking movements. At low concentrations of the extract foragers are attracted to and contact it, but at higher concentrations they are repelled by it.

In the genus *Melipona*, communication of food sources occurs without the use of pheromone trails and the function of the mandibular gland secretion appears to be related to alarm recruitment and defence. When provoked, *M. fasciata* and *M. interrupta triplaridis* rush out of the nest, land on or bite the intruding object and release their own alarm pheromone (Smith and Roubik, 1983).

Hive containing a colony of stingless bees, *Trigona augustula*, in Colombia

Workers showed the maximum response to mandibular gland extracts of their own species or to alarm gland pheromone released by a nest mate. However, the *M. fasciata* workers showed some response to glandular extracts of *M. interrupta triplaridis*. The mandibular glands of *M. fasciata* and *M. interrupta triplaridis* both contain 2-heptanol as the major constituent and substantial amounts of undecane is also present in both species; skatole and nerol are present in *M. interrupta triplaridis* only. Differences in these minor constituents could account for the differences in behavioural responses.

Workers of the genus *Lestrimelitta* lack corbiculae and attack and rob colonies of other species of stingless bee. *L. limao* workers release a pheromone secreted from their mandibular glands, that is predominantly citral, during the attack. It appears to have a conspecific recruiting and alerting function (Blum, 1966) but it it also of paramount importance in subduing the attacked colony. Its release rapidly disorientates and dispels resistance of those species of meleponine that are normally robbed by *L. limao*. In contrast species that are not robbed by *L. limao* respond mildly to citral and their defence capability is undeterred (Blum *et al.*, 1970).

Chapter fifteen

BUMBLEBEES: COLONY AND FORAGING PHEROMONES

Introduction

Compared to the honeybee, the role of bumblebee pheromones in the social communication within the colony and while foraging, has been little studied, but some interesting advances have recently been made.

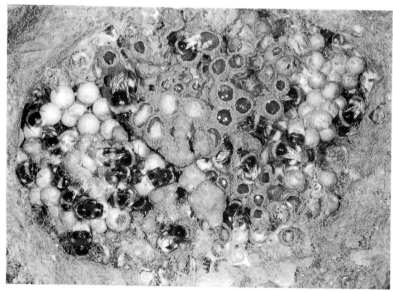

A bumblebee colony (*Bombus lucorum*)

Queen pheromone

Like honeybee workers the ovaries of bumblebee workers usually remain undeveloped, but as the season advances and the bumblebee colony becomes more populous some of the workers develop their ovaries and lay eggs. If the queen dies this happens earlier. Ovary development and egg laying is frequently associated with aggression (Free, 1955b; Free et al., 1969). Egg laying workers often jab or bite one another and are attacked by the queen if she is still present. A *Bombus lapidarius* queen tends to attack most frequently those of her workers with the most developed ovaries (Free et al., 1969). The most dominant workers in a queenless *Bombus pratorum* colony tend to spend their time astride egg cells, in which they have probably laid eggs, and try to repel intruders; a dominant worker especially attacks those whose ovary development approaches its own (Free, 1955b). In these circumstances individual recognition seems to be linked with ovary development and so possibly with the amount of pheromone produced.

It was thought that the actual behaviour of a queen bumblebee suppresses the ovary development of workers and prevents or greatly discourages them from building egg cells and laying eggs particularly during the early stages of colony growth. As a colony grows, the active dominating behaviour of the queen is no longer able to effectively keep all her large family under control (Free, 1955a; Free and Butler, 1959; van Honk et al., 1980). This very physical means of inhibition probably still holds true, but in addition the queens of at least some species are known to produce inhibitory pheromones (Röseler, 1977; Röseler and Röseler, 1978). It has been shown that in colonies of *B. terrestris*, a populous underground nesting species, queen pheromone normally hinders synthesis by the corpora allata of the worker juvenile hormone that is necessary to stimulate ovary development, although it is not completely suppressed and egg maturation proceeds, albeit very slowly. However, removal of the queen results in a high level of juvenile hormone synthesis and within five days some of the workers in a queenless colony are able to lay eggs. The process whereby the pheromone presence is transferred into a specific signal inhibiting the activity of the corpora allata is yet to be elucidated, and the queen pheromone concerned has yet to be identified.

The pheromone appears to be produced by the queen's mandibular glands and the queen probably spreads it over her body while she grooms herself. Extracts of a *B. terrestris* queen's mandibular glands and of her body surface have the same inhibitory effect as a living queen (Röseler et al., 1981). *B. terrestris* colonies headed by queens whose mandibular glands had been removed were less effective at inhibiting worker ovary development than normal queens (van Honk et al., 1980) but more effective than no queen (Pomeroy, 1981) suggesting that there is a second inhibitory source or else the mandibular gland pheromone had been incompletely removed. The pheromone appears to be very volatile because a queen that has been dead only one day is ineffective at inhibiting ovary development (Pomeroy, 1981). Indeed the presence of a dead queen can cause increased onset of worker aggression; perhaps this is because only a weak inhibitory signal is still present.

The odour alone of the queen inhibitory pheromone has some inhibitory power but for the maximum effect to be achieved the workers need to be able to make physical contact with their queen (Pomeroy, 1981). Quite what form this physical contact takes is unknown. Worker bumblebees do not form a court round their queen but they may brush against her or perhaps palpate her with their antennae. Observations are needed on this in a colony of marked bees, to determine how often individual workers contact their queen as the colony grows in adult population and in comb area. There is no decrease in the strength of the inhibitory signal of queens as they age (Pomeroy, 1981). Probably diminished contact between queens and workers is the major reason why egg laying workers occur in large colonies. Workers in large colonies would also be less likely to be subjected to the higher concentration of volatile components in the immediate vicinity of their queen. Because egg maturation can proceed at a low level even with the queen present the workers that produce eggs are likely to be the older ones. Perhaps as they age workers also become less sensitive to the inhibitory signals from the queen.

Once egg laying workers are present in the colony they are also able to inhibit egg formation in younger workers. Perhaps in achieving this the effect of their antagonistic behaviour is reinforced by pheromones they in turn produce; extracts from the mandibular glands of *B. terrestris* laying workers have some inhibitory effect (Honk & Hogeweg, 1981). The inhibitory signals from workers increase with both their age and size (Pomeroy, 1981). As a result even in large *B. terrestris* colonies the dominance by both aggressive behaviour and pheromones of the queen and a few egg laying workers maintain stability and cohesion. Probably in less advanced bumblebee species the antagonistic behaviour is of relatively more consequence and the pheromone controlling mechanism is poorly developed; removing the mandibular glands from *B. impatiens* queens makes little or no difference to their inhibitory powers (Pomeroy, 1981). Even so, pheromone production by dominant laying workers and queens probably advertises their presence to the remaining workers. Wing vibration by laying workers may help disperse the pheromone (Free, 1955b).

Pheromones from the queen may also influence queen production within the bumblebee colony. Larvae destined to become queens are fed more often than larvae reared as workers. When a *B. terrestris* colony was divided into two parts, one of which contained the foundress queen, it was discovered that larvae in the queenless part were fed more often and continued to be fed for longer (Röseler, 1970, 1974). Until late in colony development the queen pheromones may well restrain frequency of feeding so all female larvae become workers; towards the climax of colony development as the queen's influence wanes, the feeding rate may no longer be restricted and queens are produced. However, these must be regarded as tentative conclusions only and more experiments are needed.

Because bumblebees of different species show considerable differences in social structure, eventually we may well be able to identify a series of evolutionary steps on the way to complete control of reproduction by pheromones as exhibited by the honeybees.

Incubation pheromone

The presence of a brood incubation pheromone has recently been demonstrated (Heinrich, 1973). This does not appear to originate from the brood but is deposited on it. The immature stages of a bumblebee colony are incubated by the adult bees for long periods each day which help maintain them at a relatively high temperature. This incubatory behaviour appears to be released by an odour the queen herself gives to the site when she lays eggs.

The incubation pheromone also seems to guide the queen over short distances to the brood clump in the darkness of the nest, and on approaching the brood clump she palpates it with her antennae before incubating. Queens do not discriminate between incubation sites they have marked themselves and those marked by other queens of her own species. However, Heinrich (1973) found that queens of the two species he was observing (*B. vosnesenskii*

A bumblebee queen (*Bombus hortorum*) incubating her first batch of brood

and *B. edwardsii*) did not incubate brood marked by queens of the other species. When workers are present in the nest they probably also read and respond to the message but this has yet to be demonstrated; perhaps they reinforce it.

Sometimes it is possible to induce workers to rear brood of a bumblebee species other than their own, resulting in colonies populated with workers of more than one species. It would be interesting to see whether workers in such colonies learn to respond to the brood marking pheromone of the other species present.

The brood marking pheromone appears to be relatively involatile; sites and brood clumps marked by the queen remain attractive for several days after the brood has been removed. Probably a signal from the eggs initially releases brood marking behaviour, perhaps as they are laid. But the actual presence of brood is not always necessary for an incubation site to be marked as queens held in captivity sometimes incubate particular spots on the floor of their cages before they have laid any eggs.

The site at which this incubatory pheromone is produced remains to be discovered. A first step would be to mark artificially potential incubation sites with extracts from different parts of the queen's body and see which release responses.

Pheromones produced by bumblebee brood probably stimulate foraging (page 75) so it is quite possible that a pheromone from the brood itself may contribute towards the signal to incubate. No tests have been made of the relative attractiveness of two brood clumps from one of which the brood has been removed.

Trail pheromone at the nest

Nests of bumblebees are often located at the end of underground tunnels, in the midst of matted vegetation, or among an accumulation of rocks. It is likely that returning foragers would use the odour of the nest to help guide them to it, but bees leaving to forage must have some means of finding their way from the nest to the nest entrance.

When bumblebee colonies, housed in nest boxes, are given clean, glass entrance tunnels the bees initially hesitate before using them, and do not readily pass along them until they have been well used, and presumably have acquired the odour of the nest or the bees.

Cederberg (1977) demonstrated that *B. terrestris* workers both lay and follow odour trails between the nest and nest entrance. The bees were at first restricted to using a narrow path across a sheet of paper placed between nest and nest entrance. They were then given access to the whole sheet of paper but nearly all kept to the trail already established, even when it was angled at a different direction to the original. Presumably, as for honeybees (page 110), the trail pheromone is deposited by the feet and tip of the abdomen.

Production of synthetic trail pheromone of bumblebees might well prove useful in inducing bumblebee queens to discover and occupy nest boxes placed for them in the field, and so increase the numbers of bumblebee workers available for pollination.

Parasitic bees of the genus *Psithyrus* have their brood reared in bumblebee colonies. The *Psithyrus* females appear to use the bumblebee trail pheromones to recognize and locate the particular bumblebee species they parasitize (Cederberg, 1983). When *Psithyrus* females were presented with methanol extracts of the body surface of *B. lapidarius* queens they eagerly palpated them with their antennae. Extracts of all major parts of the queen, i.e. head, thorax and abdomen were attractive, but extracts of the tarsi were especially so. *P. rupestris* females readily followed odour trials of *B. lapidarius* that were either natural or made from queen extracts. They were also able to select extracts of their *B. lapidarius* host from extracts of six other bumblebee species.

Trail pheromone while foraging

Bumblebees may also mark flowers with pheromone. Cameron (1981) observed that bumblebee foragers probed artificial flowers that had a reward (honey or sucrose syrup) to offer or that had recently provided a reward, but did not probe flowers that had never offered a reward. He found that the deposit used to mark the rewarding flowers was soluble in hexane or pentane, and after artificial flowers had been washed with these solvents they no longer released probing. The extract has not yet been tested to find out whether or not it releases probing when added to non-rewarding flowers. The bumblebees gyrated and groomed themselves on rewarding flowers and this behaviour probably facilitated deposition of the pheromone. On natural flowers bumblebees make plentiful bodily contacts with petals, stamens and pistils, especially when scrabbling for pollen and when the flowers have tubular corollas.

No experiments have been done to find whether the message left by successful foragers is species specific. Attracting bees from other bumblebee colonies to favourable nectar sources could be detrimental. However, members of a bumblebee colony forage in close proximity to their nest, so probably many, if not most, of the 'messages' left on flowers would reach members of the departing bumblebee's own colony, and so help nestmates avoid flowers that are in a pre- or post- nectar secreting stage and attract them to favourable patches of flowers. Perhaps the pheromonal messages are most valuable to the departing bumblebee itself by indicating which flowers it has already visited on the current foraging trip, and which it would pay to revisit on a subsequent one.

It is likely that bumblebees, like honeybees (page 112), use a deterrent pheromone. Visits by an individual bumblebee forager to flowers on its 'foraging circuit' are in such a time sequence as to maximize the reward from each flower visit (Corbett *et al.*, 1984). Flowers probed during the past minute or so are either not landed on, or landed on but not probed. The source of such a deterrent pheromone is unknown.

For both bumblebees and honeybees the extent to which the pheromone messages are heeded must depend on the availability of forage and competition from other bees. When competition is fierce individual flowers are visited at very frequent intervals, and a bee may attempt to land on a flower while another is still present. Clearly under such conditions pheromone messages, if deposited, are ignored.

Chapter sixteen

BUMBLEBEES: MATING BEHAVIOUR

Male flight routes

One of the most fascinating aspects of bumblebee behaviour is the habit of males of many species of both *Bombus* and the parasitic genus *Psithyrus*, to fly along special routes (which may consist of a circuit up to several hundred metres long, Bringer, 1973) linking sites (e.g. a tree base or twig) which have been scented with a pheromone.

Each male has his own route, parts at least of which lies close to routes of other males of the same species and may interlock with them where the visiting sites are shared. Consequently the entire flight complex can occupy a large area. The neighbourhood of the nest itself or the nest entrance is often incorporated into the flight route (Kullenberg, 1956). Virgin queens are attracted to these networks of flight routes and so the meeting and mating of males and females is facilitated.

The first description of a flight route was made by Newman (1851):

'Any observer may watch them (*Bombus hortorum* males) in their unsteady flight, very near the ground, paying visits to the roots of trees, holes in banks etc. At first appearance they look as though they intend to alight at these haunts, but this they never do, until a round of probably a quarter of a mile is made in this manner, when they require nourishment; they then return to the thistles and flowers . . I once watched a male bee go to a spot and stop: on my reaching it, I found a queen with the drone, but they both flew away; . . .'.

A similar observation was made by Darwin (1885) on the flight paths of *B. pratorum* and *B. lucorum* males. He noted that males still continued to visit a particular site ('buzzing place') even though he had altered its visual appearance.

Scent deposition

Sladen (1912) observed that the hollows under trees and bushes that were visited by males of *B. pratorum* and *B. hortorum* acquired the same odours as the males themselves which he supposed were species specific and emanated from the males' heads. He suggested that males deposited odours at the sites and queens were attracted to them.

This suggestion was confirmed by Haas (1946) who was the first to observe that males alight at these sites and mark them with pheromone. He reported that this marking occurs early in the morning during the first or second flight of the day probably because the odour had dissipated overnight, whereas later in the day the males only approach these sites. However, later observations have shown that in variable weather males mark sites throughout the day (Svensson, 1979), and usually most males complete a few circuits or more before doing so (Awram, 1970; Awram and Free, 1987a). The same flight route area is often used from season to season, and the same objects marked with pheromone (Svensson, 1979).

Whereas the initial discovery of a route is normally by odour, once it is learned orientation to the route is visual. Removal of antennae of males that had not flown a route delayed or prevented route learning. But removal of antennae of experienced males failed to interfere with route learning. Such antennaeless, experienced males continuously scented the sites they visited; probably their inability to detect scent stimulated them to replenish it (Awram, 1970; Awram and Free, 1987a).

The male marking pheromone is produced in the cephalic part of the labial gland; the gland contents and adsorbed odour markings are chemically identical (Kullenberg *et al.*, 1973). The pheromone is secreted via a duct which opens near the base of the proboscis. It is spread over the body by grooming; consequently any object that a male touches or rests on probably acquires the marking odour. When actively scent-marking an object a male presses the ventral surface of its body against it, with its proboscis stretched backwards under its body. While scenting spruce twigs a *B. pratorum* male was seen to move its proboscis like a paint brush backwards and forwards along the needles, which acquired the odour of the marking pheromone.

Haas (1952) observed that when a male was scent marking an object such as a leaf, he held it between his mandibles and made gnawing movements. This behaviour has also been observed by Kullenberg *et al.*, (1973); its purpose has yet to be discovered. Svensson (1979) observed that *B. lapponicus* males made movements with their mandibles along the edges of leaves while use of the proboscis was particularly conspicuous when they moved along the upper sides of leaves. Awram and Free (1987a) noted that when the open mandibles were held either side of the leaf edge along which a male was progressing, the brush of hairs along the posterior edge of each mandible contacted the substrate and so may be especially important in pheromone transfer.

Usually at each site, it is possible to identify one or more objects or parts of objects which are visually conspicuous and visited by route flying males. However, when a site is being scented such a focal point of visitation receives no more attention than much of the less conspicuous background against which it is located (Awram and Free 1987a). For example the focal points for visits of *B. hortorum* males were usually bare ground between tree roots and irregularities at the base of tree trunks yet they scented leaves of plants near the bases of trees, and the leaves and bark of trees up to 3 m above ground. The focal points of visitation for *B. lucorum* and *B. terrestris* were conspicuous parts of trees, usually the stems of the branches, but they applied scent to the margins of leaves and to twigs which were often up to 2 m from the focal points.

Haas (1946, 1949a) reported that males scent marked objects located in an increasing spiral away from the site visited and in the direction of the next site. Although it is agreed that scent deposition more or less surrounds the focal point, the presence of actual scent spirals has not been confirmed by other workers (Svensson and Bergström, 1979; Awram and Free, 1987a). Indeed it is difficult to envisage how a spiral pattern could be maintained in turbulent wind conditions and most sites appear to be physically unsuitable for tracing a complex pattern. Furthermore, males fly in different directions when leaving a site so such a pattern would be superfluous as an indication of flight route.

Chemical composition of site marking pheromone

The compositions of the pheromone used by males for site marking have been studied intensively, and components of the pheromone from no less than 33 species of *Bombus* and 8 species of *Psithyrus* have now been identified (e.g. Stein, 1963; Bergström *et al.*, 1968; Calam, 1969, Kullenberg *et al.*, 1970; Svensson and Bergström, 1977 and 1979; Bergström *et al.*, 1981; Cederberg *et al.*, 1983).

The pheromones consist of mixtures of fatty acid derivatives and/or isoprenoid components (terpenes). The blend of components present is species specific and each species usually has one very dominant component, sometimes one or a few major ones, and usually several minor ones (Tables 16.1 and 16.2).

Species belonging to the same sub-species tend to show greater chemical similarity than less closely related species. Indeed the distinction between pheromones of taxonomically allied species (e.g. *B. lucorum* and *B. magnus;*

B. alpinus and *B. polaris*) sometimes appear to be based mostly on the different proportions present of the same components. The two different forms of *B. lucorum* differ in the proportions of dominant components; the main component of the dark form is ethyl dodecanoate and of the light form is ethyl tetradecenoate although each species produces both components (Bergström *et al.*, 1973). However, in contrast the marking pheromones of morphologically allied species may sometimes have a largely dissimilar chemical composition; indeed, the chemical composition has even helped to clarify the taxonomic status of the different species (Bergström, 1981).

Despite the wealth of chemical knowledge that has been made available, comparatively few ethological studies of their functions have been made, and in no case is the exact use of all the different components in a secretion known. Some components may be releases of behaviour and others may merely decrease the volatility of the releaser substances.

A very few components of cephalic gland secretions have been applied singly to leaves on flight routes and found to make them attractive (Kullenberg, 1973), but extensive investigations are needed.

Kullenberg (1973) also collected leaves scent-marked by *Psithyrus bohemicus* males and presented them elsewhere in the flight zone; males of *P. bohemicus* were attracted to them and deposited more scent in the vicinity.

In experiments conducted in large cages (Awram, 1970; Awram and Free 1987a) captive males visited plants whose leaves had been naturally scented by bumblebee males in preference to plants not scented (Table 16.3). Plants that had been artificially scented with extracts of crushed male heads were also preferred to untreated plants. The captive males also responded more to plants with male extracts of their own than of another species.

This technique should be useful as a bioassay to test the components of the site marking pheromones so far identified from male bumblebees. This must be one of the few areas of study of bee pheromones where identification of chemical components is well in advance of biological investigation of their use, and provides ideal opportunities for extensive and exciting research.

Isolating mechanisms

Specificity in site marking pheromones provides only one of a number of reproductive isolating mechanisms for a species.

Different species may establish flight routes at somewhat different periods of the season depending on the timing of their colony growth cycles but such temporal segregation is generally weak. Better segregation to prevent any attempted cross-mating is achieved by differences in location of the flight route which is often characteristic of the species concerned. For example males of some species establish flight routes in the upper parts of trees (*B. lapidarius*) and in the middle parts of trees (*B. lucorum*, *B. pascuorum*); males of other species mark and visit shrubs growing near the ground (*B. terrestris*)

Table 16.1 Distribution of marking pheromone components between 18 species belonging to three subgenera of bumble bees (after Bergström *et al.*, 1981).

Subgenus:	Bombus					Pyrobombus									Alpinobombus			
	B. lucorum (Blond)	B. lucorum (Dark)	B. terrestris	B. sporadicus	B. patagiatus	B. cingulatus	B. hypnorum	B. jonellus	B. lapponicus	B. monticola	B. pratorum	B. soroeensis	B. cullumanus	B. lapidarius	B. alpinus	B. polaris	B. hyperboreus	B. balteatus
Geraniol											×							
Citronellol											×					×		
Geranyl acetate											×							
Citronellyl acetate											×							
Farnescene isomers						×					×							
all-*trans*-Farnesol						★					★							
2, 3-Dihydro-6-*trans*-farnesol			★			★		★									+	
2, 3-Dihydro-6-*trans*-farnesal						×		+									×	
all-*trans*-Farnesyl acetate											+		×					
2, 3-Dihydrofarnesyl acetate				×														
Geranylgeraniol	×										×	×						
Geranylcitronellol		+						★	★								×	×
Geranylgeranyl acetate	×					×						+	★	★				
Dodecanol																		×
Tetradecanol						×							×					×
Hexadecadienol															×	★		
Hexadecenol						+							★		★	+		
Hexadecanol	×		×						+	+	+		+					
Octadecadienol																	×	
Octadecenol												+						★
Tetradecanal				×														
Dodecyl acetate																		×
Tetradecyl acetate			★															★
Hexadecenyl acetate										★					×			
Hexadecyl acetate	×		×						+									
Octadecyl acetate						×												
Eicosenyl acetate						×												
Eicosyl acetate		×				×												
Docosyl acetate		×																
Ethyl decanoate		×	×															
Ethyl dodecanoate	+	★	+	×	★													
Ethyl tetradecenoate	★	×	×															
Ethyl tetradecanoate	×	×																
Ethyl hexadecatrienoate	×	×																
Ethyl hexadecadienoate	×																	
Ethyl hexadecenoate	×	×				×												
Ethyl octadecatrienoate	×	×																
Ethyl octadecadienoate		×																
Ethyl octadecenoate	×	×																
Ethyl octadecanoate		×																
Dodecyl butyrate																	+	
Tetradecyl butyrate																	+	

Key ★ Main component
　　　 + Major component
　　　 × Minor component

and males of yet other species visit sites that are at ground level (*B. hypnorum*, *B. hortorum*) (Hass, 1949b; Svensson, 1979). Such physical separation is especially evident between species whose labial gland contents are chemically similar (e.g. *B. lapponicus* and *B. hypnorum*).

Cederberg *et al.*, (1983) identified the components of eight *Psithyrus* species (Table 16.2). The four species that belong to the same subgenus (*Fernaldaepsithyrus*) show the greater chemical similarity. Four of the species (*P. rupestris*, *P. bohemicus*, *P. campestris* and *P. sylvestris*) whose flight routes occupy similar spatial niches have clearly distinguishable pheromonal compositions. Two species (*P. sylvestris* and *P. quadricolor*) with pheromones of

Table 16.2   Components of labial gland secretion of eight species of *Psithyrus* (subgenera are given in brackets) (after Cederberg *et al.*, 1983).

	P. rupestris (*Psithyrus*)	*P. bohemicus* (*Ashtonipsithyrus*)	*P. campestris* (*Metapsithyrus*)	*P. barbutellus* (*Allopsithyrus*)	*P. quadricolor* (*Fernaldaepsithyrus*)	*P. sylvestris* (*Fernaldaepsithyrus*)	*P. norvegicus* (*Fernaldaepsithyrus*)	*P. flavidus* (*Fernaldaepsithyrus*)
Citronellol	+							
all-*trans*-Farnesol		★						
all-*trans*-Farnesyl acetate			+					
Geranylcitronellol	×							
Geranylgeranyl acetate	+							
Tetradecanal		×			+	+		
Tetradecanol	×							
Hexadecanal						×	×	
Hexadecenal		×				★	+	
Hexadecanol	×					×		
Hexadecenol	★	★				★	+	★
Octadecenol			+			+	×	★
Eicosenol			★					
Octadecyl acetate				×				
Ethyl tetradecanoate								+
Ethyl tetradecenoate								+
Tetradecenoic acid						+	+	

★ Main component
+ Major component
× Minor component

Table 16.3   Visits by bumblebee males to plants treated with scent (from Awram and Free 1987a).

Male bumblebee species	Plant treatment		No. of visits to plant of treatment	
	A	B	A	B
B. pratorum	scented by live B. pratorum males	untreated	57	10
B. pratorum	crushed B. pratorum male head	untreated	47	0
B. lucorum	crushed B. lucorum male head	untreated	34	1
B. terrestris	crushed B. terrestris male head	untreated	83	5
B. terrestris	crushed B. terrestris male head	crushed B. terrestris worker head	70	0
B. terrestris	crushed B. terrestris male head	crushed B. terrestris male body	44	14
B. terrestris	crushed B. terrestris male head	crushed B. lucorum male head	38	0
B. pratorum	crushed B. pratorum male head	crushed B. terrestris male head	21	0
B. pratorum	crushed B. terrestris male head	untreated	17	8

similar composition have clearly differentiated flight route locations, the former being near the ground and the latter being at tree-top level.

Lastly cross-mating is probably prevented or discouraged because a queen is unreceptive to strange males of other species or perhaps because the males themselves find her unattractive.

Queen behaviour

Free flying queens have been seen to visit the sites established by males along flight routes (e.g. Haas, 1949b; Kruger, 1951; Free, 1971; Svensson, 1979). But observations associating queens with flight routes are, unfortunately, not common, probably because the system is so efficient that the queens are quickly mated. Actual copulation probably occurs on foliage or on the ground a short distance from the visiting site.

Perhaps the male marking pheromone acts as an attractant and arrestant for the queen, but this has yet to be demonstrated experimentally. Perhaps it also acts as an aphrodisiac. The bioassay, mentioned above, to test the reactions of males to components of site marking pheromones could also be used to test their attractiveness to virgin queens.

Queens must seek out the type of location where flight routes of males of their own species occur. Because the marking pheromone desposited by males emanates from many points covering a relatively wide area, a queen is

Bumblebees (*Bombus pratorum*) mating. The male is on the left

provided with a large target to reach. It is not known whether the queen's antennae are in any way specialized for detecting male scent.

Recognition of queens

When queens are suspended by threads near the visiting sites along flight routes males are usually attracted to them and attempt to mate, but those tethered between visiting sites are usually ignored (Free, 1971). Indeed, the queen's attractiveness is greatest within a few centimetres of a focal point at the visiting site, and is greatly diminished even a short distance away (Awram, 1970; Awram and Free, 1987b).

Tethered queens and models of queens have been used to determine which features of a queen are attractive to males (Awram, 1970; Free, 1971; Awram and Free, 1987b). The initial male response is visual. The preference shown by males for models of different colours and sizes in general reflects the sizes and colour patterns of their queens, especially their appearance under ultraviolet light.

After the intial visual response, odour appears to be important in eliciting contact and attempted mating. Males were induced to alight on small cylindrical cages of perforated zinc that contained virgin queens, whereas they inspected but did not alight on similar cages containing workers or drones (Free, 1971). As visual identification was excluded in these experiments the queen odours alone must have been responsible for the males' reaction to them. When given the choice males made contact with queens of their own species more than those of other species, although the initial response to each species was similar.

Although males' initial responses to virgin and old mated queens were similar, they attempted to mate with young queens but not old ones (Free, 1971). Differences in the urge to mate could be accounted for by differences in the amount and composition of the queens' sex attractant, or some other odour change associated with age or ovary development.

Free (1971) found that the head and thorax of a queen is more attractive than the abdomen only and males rarely attempted to mate with decapitated queens. He suggested that the head was the source of any pheromones involved. It now appears probable that the sex pheromone of a queen emanates from her mandibular or labial glands. Van Honk *et al.* (1978) found that living queens whose mandibular glands had been extirpated elicited approach and inspection by males as much as did intact queens, or queens without mandibular glands but with mandibular gland secretion applied to their bodies. However, following inspection of queens without mandibular glands or mandibular gland secretion, males copulated with relatively few of them compared to the other categories. It therefore appears that the mandibular gland secretion releases mounting and copulation. None of the components of the mandibular gland secretion that are responsible have yet been identified. Nor has it been shown that they are species specific.

Hexane extracts of male heads (*B. lucorum, P. bohemicus, P. rupestris*) have been applied to 'models' of queens, consisting of small cylinders of black velvet, and found to elicit copulatory attempts by males of the species concerned (Kullenberg, 1973). Possibly therefore the male cephalic gland pheromone serves not only for terrestrial marking but also as an aphrodisiac and copulation stimulant. More investigations are necessary to differentiate between the functions and effects of the male pheromone and that from the queen's mandibular glands.

Other types of mating behaviour

The mating of species that do not have flight routes has been little studied.

Some males (*Bombus fervidus, B. latreillelus* and *B. ruderarius*) wait near the entrances of nests to mate with young queens as they emerge. Presumably they can readily differentiate between queens and workers by odour otherwise they would continuously disrupt the normal foraging flights of workers; however, this does not seem to have been studied.

Copulation sometimes occurs within the nest. But it is not known whether the males belong to the same colony as the young queens. Because most males appear to leave their parental nests when a few days old and never return the copulating males are probably visitors. Presumably the old mother queen is not attractive to them.

Males of other species (e.g. *B. confusus*) perch on various objects and dart at passing insects whose size approaches that of a bumblebee queen (Schremmer, 1972). Recognition by scent probably preceeds attempted copulation.

Chapter seventeen

CONCLUSIONS

Pheromone composition and function

It has become clear that most, if not all, bee pheromones are multi-component and communication by pheromones is much more complex than originally supposed.

Nothing is known of the chemical identity of most bee pheromones. Even pheromones that have been studied chemically contain many components that have not been identified and the function of many identified components remains to be determined; this applies especially to queen and alarm pheromones. Some of the components whose chemistry or function is unknown may be products on metabolic pathways to the main components, or could be derived from them. Some may be 'keeper' or 'holding' components of relatively high molecular weight that prolong the action of the more volatile components. But there are many still unaccounted for, no doubt the essential missing components of the queen mandibular gland pheromone are among them.

In combination with the evolution of glands that secrete and emit pheromones, bees have developed olfactory receptors of high sensitivity, capable of rapid and precise discrimination. Consequently, biologists are usually able to produce a satisfactory bioassay for use in the development of a synthetic pheromone. Lack of complete chemical identification of a pheromone is usually the prime constraint to progress in bee pheromone research. Until recently the minute quantities of material available for chemical analysis have made this task extremely difficult. Hopefully recent developments in the techniques and devices associated with the employment of the coupled gas-chromatograph and mass-spectrometer will help overcome the problem.

Many of a pheromone's components release different behaviour patterns and so have different functions or different combinations of functions. Most bioassays have investigated the components in isolation. There is less information on the ability of bees to detect and respond to a component in a mixture, and on whether the response to it is the same as when the component is presented alone. The relative proportions of the different components of a pheromone must vary at different distances from the source of secretion according to their volatility. Perhaps there is a corresponding variation in the type of response elicited.

The absence of certain components in a synthetic mixture may be partially compensated for by an increased amount of some of the others, but the correct blend is necessary to elicit the full response. It is essential to test mixtures of the different components of the same pheromone, as well as the individual components, in order to determine any additive or synergistic effects. But so far studies of bee pheromones have provided little evidence of either. The effect of two components in combination rarely seems to equal the sum of the effects of each acting on its own, and certainly has never exceeded it.

Whereas a single component of a multi-component pheromone can serve multiple functions there is no evidence yet that it ever does so unaided. The queen pheromone component 9-ODA may approach the effectiveness of the complete mandibular gland secretion in attracting drones and indeed may be the sole mandibular gland component involved, but 9-ODA alone is insufficient to induce normal court formation, and to inhibit queen rearing or ovary development. Other components of the mandibular gland secretion or other pheromones are probably always involved in these latter activities. The method of dispersal of 9-ODA may also differ in the different circumstances.

Because non-social insects communicate by pheromones during mating, this is likely to be their earliest evolutionary use by eusocial bees. It is significant that 9-ODA occurs in all honeybee species. Probably 9-ODA developed as a sex attractant and as the honeybee colony evolved it combined with additional chemical secretions to produce other signals. Perhaps its extended use was originally associated with aggressive behaviour as sometimes exemplified by honeybee and bumblebee laying workers; the capability of inhibiting reproduction by chemical means alone evolved later. Although many alarm pheromone components release more than one behaviour pattern, like 9-ODA each always normally acts only in conjunction with other components. Such a multi-functional role of queen and alarm pheromone components allows an economic use of the existing chemical repertory to produce the required output of signals.

It is tempting to suppose that bees of different physiological conditions, or with different behavioural repertoires, respond preferentially to different components of a pheromone mixture, or that different components of a pheromone may be responded to in different behavioural contexts. But there is little evidence for these suppositions. Indeed, under natural conditions the Nasonov pheromone is effective as an attractant in several different behavioural contexts (i.e. when clustering, at the hive entrance, at a source of water) but in each of these situations the nerol and farnesol components appear to have no influence or are slightly repellent.

Frequently two or more pheromones are released during a behavioural situation, sometimes from different castes. (For example from queen mandibular glands and worker Nasonov glands to attract and guide a swarm; from queen mandibular glands and tergite glands to attract a drone to mate; from worker mandibular glands and sting chambers to elicit alarm and attack, and to mark opponents; from queen mandibular glands, tarsal glands and tergite glands to advertise the queen's presence.) This must help to endorse the specificity of the signal. Furthermore, the pheromones being released will inevitably be associated with the ever present colony odour and trail pheromones.

Chemical specificity may be further aided by ecological or behavioural specialities. For example, male bumblebees mark their flight routes with highly specific attractant blends but in addition males of different species may occupy different ecological areas, fly at different heights above ground, and at somewhat different times of the season.

The confines of a honeybee colony, with numerous contacts and exchanges between individuals of the same or different castes, and between individuals at different stages of maturity must embrace numerous pheromone systems so far undiscovered. There are many possibilities. Bees performing communication dances on the comb surface may release pheromones to help complement the information from the dance itself. Submissive bees being scrutinized by guard bees may release 'appeasement' pheromones. Seeking or offering food could well be associated with special pheromones. Bees may release a pheromone signal when they are trying to induce others to groom them. Bees that are dead or dying are readily recognized and removed from the nests; they may emit a pheromone that attracts the attention of 'undertaker' bees. Special pheromones may be released prior to a swarm's departure from its mother colony, or before an entire colony migrates or absconds. The function of the Dufour gland which is so important in ground nesting solitary bees and ants could well repay study.

As more pheromone components are identified and their functions determined the sophistication and elegance of communication within the bee colony will become increasingly apparent.

Economic potential

Already it is possible to envisage the use of synthetic pheromones within the hive to control several activities that could have economic benefit.

Synthetic pheromone could be used to inhibit queen rearing and swarming; colonies could then be requeened entirely at the discretion and convenience of the beekeeper. Judicious use of synthetic queen pheromone could well be helpful in aiding acceptance of immature and mature queens by the recipient colony. It could be used to stimulate many worker activities in the nest including comb building, brood rearing, food storage and foraging and so make each colony a more efficient and productive unit. It could perhaps be used to establish mating sites in areas with desirable genetic material, and to avoid unwanted matings. A different blend of components might be necessary for each of these tasks.

Synthetic queen and brood pheromones could be used to control worker ovary development in colonies that have been queenless for many days so they more readily accept an introduced queen.

Synthetic pheromone simulating that of drone brood and adult drones could be used to prevent drone rearing. Synthetic pheromones of drone comb could be used to reduce the amount of drone comb built in the brood area of a colony. Meanwhile this can be achieved by using drone comb in the honey storage areas of hives.

The synthetic Nasonov lure, being produced commercially, is already being used successfully to attract swarms and migrating colonies to empty hives. It can also be used to encourage bees to accept artificial food and water provided inside the hive, and to use selected water sources in the field. The Nasonov lure can also help to remove stray bees from situations where their presence is undesirable; in this context it is especially effective when presented with synthetic queen pheromone. It may be useful in attracting clustering swarms directly they have left their hives, and may help prevent colonies from abandoning their hives, as frequently happens in the tropics. Finally it should be possible to use the Nasonov lure to help ensure a favourable reception from a colony of a newly introduced queen.

It has been demonstrated that adapting colonies to synthetic alarm pheromone reduces their tendency to sting. This should make it easier to requeen particularly aggressive colonies with queens of a more docile strain. This technique could prove especially useful in requeening colonies of aggressive 'Africanized' bees in South and Central America, perhaps after they have been attracted to hives baited with Nasonov pheromone. With the aid of selected synthetic components of alarm pheromone it should be possible to deter bees from robbing other colonies of their honey stores and at the same time maintain the alertness of the defending bees.

Economically the most important potential use of synthetic honeybee pheromones is to increase crop pollination (Fig. 17.1). Brood and queen pheromones stimulate foraging, so increasing the amount of them in a colony should increase its pollinating ability. If satisfactory systems can be devised of using more than one queen per colony, more foraging should result and they could be used as temporary measures until synthetic brood and queen pheromones are available.

On most crops honeybees collecting pollen are more effective as pollinators than those collecting nectar only. Increasing the amount of brood pheromone which stimulates pollen collection in particular, should also increase pollination. Until relevant synthetic brood pheromone is available minor and inexpensive adaptations to hives that direct foragers to the brood areas of colonies and so increase their exposure to natural brood pheromone can be employed. Conversely directing returning foragers away from the brood could aid in providing pollen-free combs of honey.

Many crops needing pollination are relatively unattractive to bees and even when honeybee colonies are placed at the crops only a small proportion of foragers visit them. As well as using synthetic pheromones to increase foraging and pollination, it should be possible to use them to attract bees to

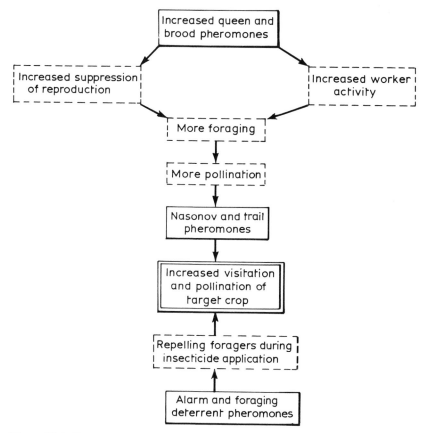

Figure 17.1 Use of pheromones to increase pollination efficiency of colonies.

the particular crops needing pollination. Although promising results have yet to be obtained it seems that an attractant based on some of the Nasonov pheromone components could be useful. Perhaps one based on the honeybee trail pheromone could be more effective.

If insecticide is applied to a crop while it is still in flower it should be possible to repel foragers from it by using synthetic alarm pheromones. Perhaps a synthetic pheromone based on that used to mark unrewarding floral sources and deter bees from visiting them could also be used in this connection.

Most of the synthetic pheromones mentioned above are far from a reality and their production is still dependent on identification of the natural material. Dispensing the synthetic pheromones within the colony may also present problems. But some synthetic pheromones are already available or soon will be. No doubt many other uses of synthetic pheromones will suggest themselves as we learn more of the many pheromonal systems employed by the honeybee colony.

Primary research objectives should be to dispense synthetic queen, brood and foraging pheromones and so programme a colony's activities to enhance crop pollination and honey production. Before this can be achieved all the chemical components of the pheromones concerned need to be identified and their functions determined so a synthetic product, equal or superior to the original can be produced. The conditions under which the pheromone is released and its mode of distribution must also be determined so that synthetic pheromones can be dispensed efficiently in the correct behavioural context.

There has already been much progress but continued intensive and prolonged research will be necessary before these targets can be achieved. Fortunately, because the honeybee is so widely distributed and valued, the results of research will have an equally widespread influence on crop and honey yields.

REFERENCES

Adler, V.E., Doolittle, R.E., Shimanuki, H. and Jacobson, M. (1973) Electrophysiological screening of queen substance and analogues for attraction to drone, queen and worker honey bees. *Journal of Economic Entomology*, **66**, 33–36.

Akratanakul, P. (1977) *The natural history of the dwarf honeybee, Apis florea F. in Thailand*. Ph.D. Thesis, Cornell University, Ithaca, New York.

Allen, M.D. (1955) Observations on honeybees attending their queen. *British Journal of Animal Behaviour*, **3**, 66–69.

Allen, M.D. (1956) The behaviour of honeybees preparing to swarm. *British Journal of Animal Behaviour*, **4**, 14–22.

Allen, M.D. (1957) Observations on honeybees examining and licking their queen. *British Journal of Animal Behaviour*, **5**, 81–84.

Allen, M.D. (1965a) The role of the queen and males in the social organisation of insect communites. *Symposia of the Zoological Society of London*, **No. 14**, 133–57.

Allen, M.D. (1965b) The production of queen cups and queen cells in relation to the general development of honeybee colonies and its connection with swarming and supersedure. *Journal of Apicultural Research*, **4**, 121–41.

Al-Sa'ad, B.N., Free, J.B. and Howse, P.E. (1985) Adaptation of worker honeybees (*Apis mellifera*) to their alarm pheromones. *Physiological Entomology*, **10**, 1–14.

Al-Sa'ad, B.N., Free, J.B., Howse, P.E., Ferguson, A.W. and Simpkins, J.R. (1987) The ability of components of the honeybee alarm pheromone to release stinging. (Not yet published.)

Al-Tikrity, W.S., Benton, A.W., Hillman, R.C. and Clarke, Jr, W.W. (1972) The relationship between the amount of unsealed brood in honeybee colonies and their pollen collection. *Journal of Apicultural Research*, **11**, 9–12.

Ambrose, J.T. (1975) Aggressive behaviour of honeybee workers towards foreign queens and an inconsistency with the stress-pheromone hypothesis. *Canadian Journal of Zoology*, **53**, 69–71.

Ambrose, J.T., Morse, R.A. and Boch, R. (1979) Queen discrimination by honey-bee swarms. *Annals of the Entomological Society of America*, **72**, 673–75.

Avitabile, A. (1978) Brood rearing in honeybee colonies from late autumn to early spring. *Journal of Apicultural Research*, **17**, 69–73.

Avitabile, A., Morse, R.A. and Boch, R. (1975) Swarming honey bees guided by pheromones. *Annals of the Entomological Society of America*, **68**, 1079–82.

Awram, W.J. (1970) Flight route behaviour of bumblebees, PhD Thesis, University of London.

Awram, W.J. and Free, J.B. (1987a) Observations on the flight routes of bumblebee males and the development of a bioassay for site marking pheromones. (Not yet published.)

Awram, W.J. and Free, J.B. (1987b) Reactions of male bumblebees (*Bombus* spp.) to queens on their flight routes. (Not yet published.)

Bai, A.R.K. and Reddy, C.C. (1975) Ovary development and egg laying in *Apis cerana indica* workers. *Journal of Apicultural Research*, **14**, 149–52.

Baird, D.H. and Seeley, T.D. (1983) An equilibrium theory of queen production in honeybee colonies preparing to swarm. *Behavioural Ecology & Sociobiology*, **13**, 221–28.

Barbier, J. and Lederer, E. (1960) Structure chimique de la substance royale de la reine d'abeille (*Apis mellifica* L.). *Comptes Rendus des Séances de L'Académie de Science Paris*, **251**, 1131–35.

Barbier, M., Lederer, E. and Nomura, T. (1960) Synthèse de l'acide céto-9-décène-2-trans-oïque (substance royale) et de l'acide céto-8-nonène-2-trans oïque. *Comptes Rendus des Séances de L'Académie des Sciences (Paris)*, **251**, 1131–35.

Barbier, M. and Pain, J. (1960) Study of the secretion of queen and worker mandibular glands (*Apis mellifica*) by gas-phase chromatography. *Comptes rendus des Séances de L'Académie des Sciences*, **250**, 3740–42.

Baribeau, M. (1976) Timing on introducing queen cells when requeening established colonies. *American Bee Journal*, **116**, 109.

Barker, R.J. (1960) Personal communication quoted by Gary (1961b).

Barker, R.J. (1971) The influence of food inside the hive on pollen collection by a honeybee colony. *Journal of Apicultural Research*, **10**, 23–26.

Barker, R.G. and Jay, S.C. (1974) A comparison of foraging activity of honey bee colonies with large and small populations. *The Manitoba Entomologist*, **8**, 48–54.

Barrows, E.M., Bell, W.J. and Michener, C.D. (1975) Individual odour differences and their social functions in insects. *Proceedings of the National Academy of Science*, **72**, 2824–28.

Beetsma, J. and Schoonhoven, L.M. (1966) Some chemosensory aspects of the social relations between the queen and the worker in the honeybee (*Apis mellifera* L.) *Proceedings Koninklijke Nederlandse Akademie van Wetenschappen, Series C*, **69**, 645–47.

Bergström, G. (1981) Chemical aspects of insect exocrine signals as a means for systematic and phylogenetic discussions in aculeate hymenoptera. *Entomologica Scandinavica. No. 15, Supplement*, 173–84.

Bergström, G., Kullenberg, B., Ställberg-Stenhagen, S. and Stenhagen, E. (1968) Studies on natural odoriferous compounds. II. Identification of a 2,3-dihydro-farnesol as the main component of the marking perfume of male bumblebees of the species *Bombus terrestris*. *Arkiv För Kemi*, **28**, 453–69.

Bergström, G., Kullenberg, B. and Ställberg-Stenhagen, S. (1973) Studies on natural odoriforous compounds. VII Recognition of two forms of *Bombus lucorum* L. (Hymenoptera, Apidae) by analysis of the volatile marking-secretion from individual males. *Chemica Scripta*, **4**, 174–82.

Bergström, G., Svensson, B.G., Appelgren, M. and Groth, I. (1981) Complexity of bumble bee marking pheromones: biochemical, ecological and systematical interpretations. In *The Systematics Association Special Volume No. 19: Biosystematics of Social Insects* (eds P.E. Howse and J.L. Clément), pp. 175–83.

Blum, M.S. (1966) Chemical releasers of social behaviour. VIII Citral in the mandibular gland secretion of *Lestrimelitta limdo*. *Annals of the Entomological Society of America*, **59**, 962–64.

Blum, M.S. (1971) The chemical basis of insect sociality. In *Chemicals Controlling Insect Behaviour*. (ed. M. Beroza) Academic Press, London, pp. 61–94.

Blum, M.S. (1974) Pheromonal bases of social manifestations in insects. In *Pheromones* (ed. M.C. Birch), North Holland/Elsevier, Amsterdam. pp. 190–99.

Blum, M.S. (1977) Pheromonal communication in social and semisocial insects. In *Proceedings of a Symposium on Insect Pheromones and Their Applications*, Nagaoka and Tokyo, December 8–11, 1976. Tokyo, Japan; Japan Plant Protection Association (1977) pp. 49–60.

Blum, M.S. (1982) Pheromonal bases of insect sociality: communications, conundrums and caveats. In *Les Médiateurs Chimiques*, Versailles, 1981. (ed. INRA Publications).

Blum, M. S. (1984) Personal communication.

Blum, M.S., Crewe, R.M., Kerr, W.E., Keith, L.H., Garrison, A.W. and Walker, M.M. (1970) Citral in stingless bees: Isolation and functions in trail-laying and robbing. *Journal of Insect Physiology*, **16**, 1637–48.

Blum, M.S., Boch, R., Doolittle, E., Tribble, M.T. and Traynham, J.G. (1971) Honey bee sex attractant: conformational analysis, structural specificity, and lack of masking activity of congeners. *Journal of Insect Physiology*, **17**, 349–64.

Blum, M.S., Fales, H.M., Tucker, K.W. and Collins, A.M. (1978) Chemistry of the sting apparatus of the worker honeybee. *Journal of Apicultural Research*, **17**, 218–21.

Boch, R. (1979) Queen substance pheromone produced by immature queen honeybees. *Journal of Apicultural Research*, **18**, 12–15.

Boch, R and Avitabile, A. (1979) Requeening honeybee colonies without dequeening. *Journal of Apicultural Research*, **18**, 47–51.

Boch, R. and Lensky, Y (1976) Pheromonal control of queen rearing in honeybee colonies. *Journal of Apicultural Research*, **15**, 59–62.

Boch, R. and Morse, R.A. (1974) Discrimination of familiar and foreign queens by honey bee swarms. *Annals of the Entomological Society of America*, **67**, 709–11.

Boch, R. and Morse, R.A. (1979) Individual recognition of queens by honey bee swarms. *Annals of the Entomological Society of America*, **72**, 51–53.

Boch, R. and Morse, R.A. (1981) Effects of artificial odours and pheromones on queen discrimination by honeybees (*Apis mellifera* L.) *Annals of the Entomological Society of America*, **74**, 66–67.

Boch, R. and Morse, R.A. (1982) Genetic factor in queen recognition odours of honey bees. *Annals of the Entomological Society of America*, **75**, 654–56.

Boch, R. and Rothenbuhler, W.C. (1974) Defensive behaviour and production of alarm pheromone in honeybees. *Journal of Apicultural Research*, **13**, 217–21.

Boch, R. and Shearer, D.A. (1962) Identification of geraniol as the active component in the Nassonoff pheromone of the honeybee. *Nature*, **194**, 704–06.

Boch, R. and Shearer, D.A. (1963) Production of geraniol by honey bees of various ages. *Journal of Insect Physiology*, **9**, 431–34.

Boch, R. and Shearer, D.A. (1964) Identification of nerolic and geranic acids in the Nassonoff pheromone of the honey bee. *Nature*, **202**, 320–21.

Boch, R. and Shearer, D.A. (1965a) Attracting honeybees to crops which require pollination. *Americal Bee Journal*, **105**, 166–67.

Boch, R. and Shearer, D.A. (1965b) Alarm in the beehive. *American Bee Journal*, **105**, 206–07.

Boch, R. and Shearer, D.A. (1966) Iso-pentyl acetate in stings of honeybees of different ages. *Journal of Apicultural Research*, **5**, 65–70.

Boch, R. and Shearer, D.A. (1967) 2-heptanone and 10-hydroxy-trans-dec-2-enoic acid in the mandibular glands of worker honey bees of different ages. *Zeitschrift fur Vergleichende Physiologie*, **54**, 1–11.

Boch, R. and Shearer, D.A. (1971) Chemical releasers of alarm behaviour in the honey bee, *Apis mellifera. Journal of Insect Physiology*, **17**, 2277–85.

Boch, R., Shearer, D.A. and Stone, B.C. (1962) Identification of iso-amyl acetate as an active component in the sting pheromone of the honeybees. *Nature*, **195**, 1018–20.

Boch, R., Shearer, D.A. and Petrasovits, A. (1970) Efficacies of two alarm substances of the honey bee. *Journal of Insect Physiology*, **16**, 17–24.

Boch, R., Shearer, D.A. and Young, J.C. (1975) Honeybee pheromones: field tests of natural and artificial queen substance. *Journal of Chemical Ecology*, **1**, 133–48.

Boeckh, J., Kaissling, K.E. and Schneider, D. (1965) Insect olfactory receptors. *Cold Spring Harbour Symposia of Quantitative Biology*, **30**, 263–80.

Breed, M.D. (1981) Individual recognition and learning of queen odours by worker honeybees. *Proceedings of the National Academy of Science*, **78**, 2635–37.

Breed, M.D. (1983) Nestmate recognition in honey bees. *Animal Behaviour*, **31**, 86–91.

Breed,M.D. and Gamboa, G.J. (1977) Behavioural control of workers by queens in primitively eusocial bees. *Science*, **195**, 694–96.

Bringer, B. (1973) Territorial flight of bumble-bee males in coniferous forest on the northernmost part of the Island of Öland. *Zoon* **1**, 15–22.

Buckle, G.R. and Greenberg, L. (1981) Nestmate recognition in sweat bees (*Lassioglossum zephyrum*): Does an individual recognise its own odour or only odours of its nestmates? *Animal Behaviour*, **29**, 802–09.

Butler, C. (1609) The feminine monarchie, or a treatise concerning bees. Joseph Barnes, Oxford.

Butler, C.G. (1954a) The method and importance of the recognition by a colony of honeybees (*A. mellifera*) of the presence of its queen. *Transactions of the Royal Entomological Society of London*, **105**, 11–29.

Butler, C.G. (1954b) The importance of 'queen substance' in the life of a honeybee colony. *Bee World*, **35**, 169–76.

Butler, C.G. (1954c) The World of the Honeybee. Collins, London.

Butler, C.G. (1956) Some further observations on the nature of 'queen substance' and its role in the organisation of a honey-bee (*Apis mellifera*) community. *Proceedings of the Royal Entomological Society, London*, **31**, 12–16.

Butler, C.G. (1957a) The process of queen supersedure in colonies of honeybees (*Apis mellifera* Linn.). *Insectes Sociaux*, **4**, 211–23.

Butler, C.G. (1957b) The control of ovary development in worker honeybees (*Apis mellifera*). *Experientia*, **13**, 1–6.

Butler, C.G. (1959) The source of the substance produced by a queen honeybee (*Apis mellifera*) which inhibits development of the ovaries of the workers of her colony. *Proceedings of the Royal Entomological Society of London*, **34**, 137–38.

Butler, C.G. (1960a) The significance of queen substance in swarming and supersedure in honeybee (*Apis mellifera* L.) colonies. *Proceedings of the Royal Entomological Society of London*, Series A, **35**, 129–32.

Butler, C.G. (1960b) Queen substance production by virgin queen honeybees, (*Apis mellifera*). *Proceedings of the Royal Entomological Society, London*, **35**, 170–71.

Butler, C.G. (1960c) Queen recognition by worker honeybees (*Apis mellifera* L.) *Experientia* **16**, 424–26.

Butler, C.G. (1961) The scent of queen honeybees (*A. mellifera*) that causes partial inhibition of queen rearing. *Journal of Insect Physiology*, **7**, 258–64.

Butler, C.G. (1966a) Die Wirkung von Königinnen – Extrakten verschiedener sozialer Insekten auf die Aufzucht von Königinnen und die Entwicklung der Ovarien von Arbeiterinnen der Honigbiene (*Apis mellifera*). *Zeitschrift für Bienenforschung*, **8**, 143–·47.

Butler, C.G. (1966b) Mandibular gland pheromone of worker honey bee. *Nature*, **212**, 530.

Butler, C.G. (1967) Insect pheromones. *Biological Research*, **42**, 42–87.

Butler, C.G. (1969) Some pheromones controlling honeybee behaviour. *Proceedings of the VI Congress of the International Union for the Study of Social Insects*. Berne. pp. 19–32.

Butler, C.G. (1971) The mating behaviour of the honeybee (*Apis mellifera* L.). *Journal of Entomology*, **46**, 1–11.

Butler, C.G. (1973) The queen and the "Spirit of the Hive". *Proceedings of the Royal Entomological Society of London*, Series A. **48**, 59–65.

Butler, C.G. and Calam, D.H. (1969) Pheromones of the honey bee – the secretion of the Nassonoff gland of the worker. *Journal of Insect Physiology*, **15**, 237–44.

Butler, C.G. and Callow, R.K. (1968) Pheromones of the honeybee (*Apis mellifera* L.): the 'inhibitory scent' of the queen. *Proceedings of the Royal Entomological Society of London*, **43**, 62–65.

Butler, C.G. and Fairey, E.M. (1963) The role of the queen in preventing oogenesis in worker honeybees. *Journal of Apicultural Research*, **2**, 14–18.

Butler, C.G. and Fairey, E.M. (1964) Pheromones of the honeybee: biological studies of the mandibular gland secretion of the queen. *Journal of Apicultural Research*, **3**, 65–76.

Butler, C.G. and Free, J.B. (1952) The behaviour of worker honeybees at the hive entrance. *Behaviour*, **4**, 262–84.

Butler, C.G. and Gibbons, D.A. (1958) The inhibition of queen rearing by feeding queenless worker honeybees (*A. mellifera*) with an extract of queen substance. *Journal of Insect Physiology*, **2**, 61–64.

Butler, C.G. and Paton, P.N. (1962) inhibition of queen rearing by queen honey bees (*Apis mellifera* L.) of different ages. *Proceedings of the Royal Entomological Society, London* (A), **37**, 114–16.

Butler, C.G. and Simpson, J. (1956) The introduction of virgin and mated queens, directly and in a simple cage. *Bee World*, **37**, 105–14.

Butler, C.G. and Simpson, J. (1958) The source of the queen substance of the honeybee (*Apis mellifera* L.). *Proceedings of the Royal Entomological Society, London* (A), **33**, 120–22.

Butler, C.G. and Simpson, J. (1965) Pheromones of the honeybee (*Apis mellifera* L.) an olfactory pheromone from the Koschewnikow gland of the queen. *Vědecké Práce, Scientific Studies, University of Libcice, Czechoslovakia*, **4**, 33–6.

Butler, C.G. and Simpson, J. (1967) Pheromones of the queen honeybee (*Apis mellifera* L.) which enable her workers to follow her when swarming. *Proceedings of the Royal Entomological Society, London*, **42**, 149–54.

Butler, C.G., Callow, R.K. and Johnson, N.C. (1961) The isolation and synthesis of queen substance, 9-oxodec-trans-2-enoic acid, a honeybee pheromone. *Proceedings of the Royal Entomological Society*, **155**, 417–32.

Butler, C.G., Calam, D.H. and Callow, R.K. (1967) Attraction of *Apis mellifera* drones by the odours of the queens of two other species of honeybees. *Nature*, **213**, 423–24.

Butler, C.G., Fletcher, D.J.C. and Watler, D. (1969) Nest-entrance marking with pheromones by the honeybee – *Apis mellifera* L., and by a wasp, *Vespula vulgaris* L. *Animal Behaviour*, **17**, 142–47.

Butler, C.G., Fletcher, D.J.C. and Watler, D. (1970) Hive entrance finding by honeybee (*Apis mellifera*) foragers. *Animal Behaviour*, **18**, 78–91.

Butler, C.G., Free, J.B. and Williams, I.H. (1971) Unpublished.

Butler, C.G., Callow, R.K. Koster, C.G. and Simpson, J. (1973) Perception of the queens by workers in the honeybee colony. *Journal of Apicultural Research*, **12**, 159–66.

Butler, C.G., Callow, R.K., Greenway, A.R. and Simpson, J. (1974) Movement of the pheromone, 9-oxodec-2-enoic acid, applied to the body surfaces of honeybees (*Apis mellifera*). *Entomologi a experimentalis et applicata*, **17**, 112–16.

Calam, D.H. (1969) Species and sex-specific compounds from the heads of male bumblebees (*Bombus* spp.). *Nature*, **221**, 856–57.

Callow, R.K. and Johnston, N.C. (1960) The chemical constitution and synthesis of queen substances of honeybees (*Apis mellifera* L.). *Bee World*, **41**, 152–53.

Callow, R.K. Johnston, N.C. and Simpson, J. (1959) 10-hydroxy-Δ^2-decenoic acid in the honeybee (*Apis mellifera*). *Experientia*, **15**, 421.

Callow, R.K., Chapman, J.R. and Paton, P.N. (1964) Pheromones of the honeybee: Chemical studies of the mandibular gland secretion of the queen. *Journal of Apicultural Research*, **3**, 77–89.

Cameron, S.A. (1981) Chemical signals in bumble bee foraging. *Behavioural Ecology and Sociobiology*, **9**, 257–60.

Caron, D.M. (1979) Queen cup and queen cell production in honeybee colonies. *Journal of Apicultural Research*, **18**, 253–56.

Cederberg, B. (1977) Evidence for trail marking in *Bombus terrestris* workers (Hymenoptera, Apidae) *Zoon*, **5**, 143–46.

Cederberg, B. (1983) The role of trail pheromones in host selection by *Psithyrus rupestris* (Hymenoptera, Apidae). *Annales Entomologiska Fennica*, **49**, 11–16.

Cederberg. B., Svensson, Bo. G., Bergstrom, G., Appelgren, M. and Groth, I. (1983) Male marking pheromones in north European cuckoo bumble bees, *Psithyrus* (Hymenoptera, Apidae). *Nova Acta Regiae Societatis Scientiarum Upsaliensis*, **3**, 161–66.

Chaudhry, M. and Johansen, C.A. (1971) Management practices affecting efficiency of the honey bee, *Apis mellifera*. (Hymenoptera: Apidae). *Melanderia*, **6**, 1–32.

Chauvin, R. (1962) Sur l'épagine et sur les glandes tarsales d'Arnhart. *Insectes Sociaux*, **9**, 1–5.

Chauvin, R., Darchen, R. and Pain, J. (1961) Sur l'existence d'une hormone de construction chez les abeilles. *Comptes Rendus des Séances de L'Académie de Science*, **253**, 1135–36.

Collins, A.M. (1980) Effect of age on the response to alarm pheromones by caged honeybees. *Annals of the Entomological Society of America*, **73**, 307–09.

Collins, A.M. (1981) Effects of temperature and humidity on honeybee response to alarm pheromones. *Journal of Apicultural Research*, **20**, 13–18.

Collins, A.M. and Blum, M.S. (1982) Bioassay of compounds derived from the honeybee sting. *Journal of Chemical Ecology*, **8**, 463–70.

Collings, A.M. and Blum, M.S. (1983) Alarm responses caused by newly identified compounds derived from the honeybee sting. *Journal of Chemical Ecology*, **9**, 57–65.

Collins, A.M. and Kubasek, K.J. (1982) Field test of honey bee (Hymenoptera: Apidae) colony defensive behaviour. *Annals of the Entomological Society of America*, **75**, 383–87.

Collins, A.M. and Rinderer, T.E. (1985) Effect of empty comb on defensive behaviour of honeybees. *Journal of Chemical Ecology*, **11**, 333–38.

Collins, A.M. and Rothenbuhler, W.C. (1978) Laboratory test of the response to an alarm chemical, isopentyl acetate, by *Apis mellifera*. *Annals of the Entomological Society of America*, **71**, 906–09.

Collins, A.M., Rinderer, T.E., Harbo, J.R. and Bolten, A.B. (1982). Colony defense by Africanised and European honeybees. *Science*, **218**, 72–74.

Corbet, S.A., Kerslake, C.J.C., Brown, D. and Morland, W.E. (1984) Can bees select nectar-rich flowers in a patch? *Journal of Apicultural Research*, **23**, 234–42.

Crewe, R.M. (1982) Compositional variability: The key to the social signals produced by honeybee mandibular glands. Reprinted from: *The Biology of Social Insects. Proceedings of the 9th Congress of the International Union for the Study of Social Insects.* Boulder, Colorado, 318–22.

Crewe, R.M. and Hastings, H. (1976) Production of pheromones by workers of *Apis mellifera adansonii. Journal of Apicultural Research*, **15**, 149–54.

Crewe, R.M. and Velthuis, H.H.W. (1980) False queens: a consequence of mandibular gland signals in worker honeybees. *Naturwissenschaften*, **67**, 467.

Cruz-Landim, C. da, and Rodriquez, L. (1967) Comparative anatomy and histology of alimentary canal of adult *Apinae. Journal of Apicultural Research*, **6**, 17–28.

Cruz-Landim, C. da, and Ferreira, A. (1968) Mandibular gland development and communication in field bees of *Trigona (Scaplotrigona) postica. Journal of the Kansas Entomological Society*, **41**, 474–81.

Darchen, R. (1957) La reine d'*Apis mellifica*, les ouvrières pondeuses et les constructions cirières. *Insectes Sociaux*, **4**, 321–5.

Darchen, R. (1968) Le rôle de la reine dans la construction cirière. *Études en nucleus du nombre limite; des conséquences de la presence ou de l'absence de la reine. Traité de Biologie de L'Abeille Tome II.* (ed. R. Chauvin) Masson et Cie, Paris, pp. 305–24.

Darchen, R., Vizier, C. and Vuillaume, M. (1957) To determine the construction of drone cells in the hive. *Comptes Rendus des Séances de L'Académie de Science*, **244**, 391–4.

Darwin, C. (1885) Über die Wege der Hummelmännchen. Cited by: Freeman, R.B. (1968) *Bulletin of the British Museum of Natural History*, (Historical Series), **3**, 177–89.

Doolittle, R.E., Blum, M.S. and Boch, R. (1970) Cis-9-oxo-2-decenoic acid; synthesis and evaluation as a honey bee pheromone and masking agent. *Annals of the Entomological Society of America*, **63**, 1181–85.

Dreher, K. (1936) Die Funktion der Mandibeldrüsen bei der honigbiene (*Apis mellifica* L.). *Zoologische Anzeiger*, **113**, 26–28.

Ebadi, R. and Gary, N.E. (1980) Acceptance by honeybee colonies of larvae in artificial queen cells. *Journal of Apicultural Research*, **19**, 127–32.

van Erp, A. (1960) Mode of action of the inhibitory substance of the honeybee queen. *Insectes Sociaux*, **7**, 207–11.

Farrar, C.L. (1932) The influence of the colony's strength on brood rearing. *Annual Report of the Beekeeper's Association, Ontario, 1930 & 1931*, Ontario Department of Agriculture, pp. 126–30.

Farrar, C.L. (1937) The influence of colony populations on honey production. *Journal of Agricultural Research*, **54**, 945–54.

Farrar, C.L. (1958) *Two-queen colony management for production of honey.* United States Department of Agriculture publication, No. ARS-33-48, 11, pp.

Ferguson, A.W. and Free, J.B. (1979) Production of a forage-marking pheromone by the honeybee. *Journal of Apicultural Research*, **18**, 128–35.

Ferguson, A.W. and Free, J.B. (1980) Queen pheromone transfer within honeybee colonies. *Physiological Entomology*, **5**, 359–66.

Ferguson, A.W. and Free, J.B. (1981) Factors determining the release of Nasonov pheromone by honeybees at the hive entrance. *Physiological Entomology*, **6**, 15–19.

Ferguson, A.W., Free, J.B., Pickett, J.A. and Winder, M. (1979) Techniques for studying honeybee pheromones involved in clustering and experiments on the effect of Nasonov and queen pheromones. *Physiological Entomology*, **4**, 339–44.

Forster, I.W. (1974) Recent work by research division into various aspects of practical beekeeping. *Proceedings of the 1974 Beekeepers Seminar, Taupo* (ed. G.M. Walton) published by New Zealand Ministry of Agriculture and Fisheries, Palmerston North, pp. 126–34.

Free, J.B. (1954) The behaviour of robber honeybees. *Behaviour*, **7**, 233–41.

Free, J.B. (1955a) The division of labour within bumblebee colonies. *Insectes Sociaux*, **2**, 195–212.

Free, J.B. (1955b) The behaviour of egg-laying workers of bumblebee colonies. *British Journal of Animal Behaviour*, **3**, 147–53.

Free, J.B. (1956) A study of the stimuli which release the food begging and offering responses of worker honeybees. *British Journal of Animal Behaviour*, **4**, 94–101.

Free, J.B. (1957) The transmission of food between worker honeybees. *British Journal of Animal behaviour*, **5**, 41–7.

Free, J.B. (1958) The drifting of honeybees. *Journal of Agricultural Science*, **51**, 294–306.

Free, J.B. (1961a) Hypopharyngeal gland development and division of labour in honeybee (*Apis mellifera* L.) colonies. *Proceedings of the Royal Entomological Society, London*, **36**, 5–8.

Free, J.B. (1961b) The stimuli releasing the stinging response of honeybees. *Animal Behaviour*, **9**, 193–96.

Free, J.B. (1962) The attractiveness of geraniol to foraging honeybees. *Journal of Apicultural Research*, **1**, 52–4.

Free, J.B. (1967a) The production of drone comb by honeybee colonies. *Journal of Apicultural Research*, **6**, 29–36.

Free, J.B. (1967b) Factors determining the collection of pollen by honeybee foragers. *Animal Behaviour*, **15**, 134–44.

Free, J.B. (1968a) The conditions under which foraging honeybees expose their Nasonov gland. *Journal of Apicultural Research*, **7**, 139–45.

Free, J.B. (1968b) Engorging of honey by worker honeybees when their colony is smoked. *Journal of Apicultural Research*, **7**, 135–8.

Free, J.B. (1970a) *Insect pollination of crops*. Academic Press, London

Free, J.B. (1970b) Effect of flower shapes and nectar guides on the behaviour of foraging honeybees. *Behaviour*, **37**, 269–85.

Free, J.B. (1971) Stimuli eliciting mating behaviour of bumblebee (*Bombus pratorum* L.) males. *Behaviour*, **40**, 55–61.

Free, J.B. (1978) Progress towards the use of pheromones to stimulate pollination by honeybees. *Proceedings of the Fourth Symposium on Pollination, Maryland Agricultural Experimental Station, Special Miscellaneous Publications*, **1**, 7–22.

Free, J.B. (1979) Managing honeybee colonies to enhance the pollen-gathering stimulus from brood pheromones. *Applied Animal Ethology*, **5**, 173–78.

Free, J.B. (1981) unpublished.

Free, J.B. and Butler, C.G. (1955) An analysis of the factors involved in the formation of a cluster of honeybees (*Apis mellifera*). *Behaviour*, **7**, 304–16.

Free, J.B. and Butler, C.G. (1958) The size of apertures through which worker honeybees will feed one another. *Bee World*, **39**, 40–42.

Free, J.B. and Butler, C.G. (1959) *Bumblebees*. London, Collins.

Free, J.B. and Ferguson, A.W. (1978) Unpublished.

Free, J.B. and Ferguson, A.W. (1982) Transfer of pheromones from immature queen honeybees *Apis mellifera*. *Physiological Entomology*, **7**, 401–406.

Free, J.B. and Preece, D.A. (1969) The effect of the size of a honeybee colony on its foraging activity. *Insectes Sociaux*, **16**, 73–8.

Free, J.B. and Racey, P.A. (1966) The pollination of *Freesia refracta* in glasshouses, *Journal of Apicultural Research*, **5**, 177–82.

Free, J.B. and Racey, P.A. (1968) The effect of the size of honeybee colonies on food consumption, brood rearing and the longevity of the bees during winter. *Entomologia Experimentalis et Applicata*, **11**, 241–49.

Free, J.B. and Simpson, J. (1968) The alerting pheromones of the honeybee. *Zeitschrift fur Vergleichende Physiologie*, **61**, 361–65.

Free, J.B. and Spencer-Booth, Y. (1961) Analysis of honey farmers' records on queen rearing and queen introduction. *Journal of Agricultural Science*, **56**, 325–31.

Free, J.B. and Spencer-Booth, Y. (1965) Unpublished.

Free, J.B. and Williams, I.H. (1970) Exposure of Nasonov gland by honeybees (*Apis mellifera*) collecting water. *Behaviour*, **37**, 286–90.

Free, J.B. and Williams, I.H. (1971) Unpublished.

Free, J.B. and Williams, I.H. (1972) Hoarding by honeybees (*Apis mellifera* L.) *Animal Behaviour*, **20**, 327–34.

Free, J.B. and Williams, I.H. (1973) Unpublished.

Free, J.B. and Williams, I.H. (1974) Factors determining food storage and brood rearing in honeybee (*Apis mellifera* L.) comb. *Journal of Entomology*, Series A, **49**, 47–63.

Free, J.B. and Williams, I.H. (1975) Factors determining the rearing and rejection of drones by the honeybee colony. *Animal Behaviour*, **23**, 650–75.

Free, J.B. and Williams, I.H. (1976) The effect on the foraging behaviour of honeybees of the relative locations of the hive entrance and brood combs. *Applied Animal Ethology*, **2**, 141–54.

Free, J.B. and Williams, I.H. (1979) Communication by pheromones and other means in *Apis florea* colonies. *Journal of Apicultural Research*, **18**, 16–25.

Free, J.B. and Williams, I.H. (1983) Scent-marking of flowers by honeybees. *Journal of Apicultural Research*, **22**, 86–90.

Free, J.B. and Winder, M.E. (1983) Brood recognition by honeybee (*Apis mellifera*) workers. *Animal Behaviour*, **31**, 539–45.

Free, J.B., Weinberg, I. and Whiten, A. (1969) The egg-eating behaviour of *Bombus lapidarius* L. *Behaviour*, **35**, 313–17.

Free, J.B., Ferguson, A.W. and Pickett, J.A. (1981a) Evaluation of the various components of the Nasonov pheromone used by clustering honeybees. *Physiological Entomology*, **6**, 263–68.

Free, J.B., Pickett, J.A. Ferguson, A.W. and Smith, M.C. (1981b) Synthetic pheromones to attract honeybee (*Apis mellifera*) swarms. *Journal of Agricultural Science*, **97**, 427–31.

Free, J.B., Ferguson, A.W., Pickett, J.A. and Williams, I.H. (1982a) Use of unpurified Nasonov pheromone components to attract clustering honeybees. *Journal of Apicultural Research*, **21**, 26–9.

Free, J.B ., Williams, I.H., Pickett, J.A., Ferguson, A.W. and Martin, A. (1982b) Attractiveness of (Z)-11-eicosen-1-ol to foraging honeybees *Apis mellifera* L. (Hymenoptera, Apidae). *Journal of Apicultural Research*, **21**, 151–6.

Free, J.B., Ferguson, A.W. and Pickett, J.A. (1983a) A synthetic pheromone lure to induce worker honeybees to consume water and artificial forage. *Journal of Apicultural Research*, **22**, 224–28.

Free, J.B., Ferguson, A.W., Simpkins, J.R. and Al-Sa'ad, B.N. (1983b) Effect of honeybee Nasonov and alarm pheromone components on behaviour at the nest entrance. *Journal of Apicultural Research*, **22**, 214–23.

Free, J.B., Ferguson, A.W. and Simpkins, J.R. (1984a) Influence of immature honeybees (*Apis mellifera*) on queen rearing and foraging. *Physiological Entomology*, **9**, 387–94.

Free, J.B., Pickett, J.A., Ferguson, A.W., Simpkins, J.R. and Williams, C. (1984b) Honeybee Nasonov pheromone lure. *Bee World*, **65**, 175–81.

Free, J.B., Ferguson, A.W. and Simpkins, J.R. (1984c) Synthetic pheromone lure, useful for trapping stray honeybees. *Journal of Apicultural Research*, **23**, 88–9.

Free, J.B., Ferguson, A.W. and Simpkins, J.R. (1985a) Influence of virgin honeybees (*Apis mellifera*) on queen rearing and foraging. *Physiological Entomology*, **10**, 271–4.

Free, J.B., Pickett, J.A., Ferguson, A.W., Simpkins, J.R. and Smith, M.C. (1985b) Repelling foraging honeybees with alarm pheromone. *Journal of Agricultural Science*, **105**, 255–60.

Free, J.B., Ferguson, A.W. and Simpkins, J. (1987a) The behaviour of queen honeybees (*Apis mellifera*) and their attendants. (Not yet published).

Free, J.B., Ferguson, A.W. and Simpkins, J.R. (1987b) The effect of mature and immature queens on stimulating comb building. (Not yet published).

Free, J.B., Ferguson, A.W. and Simpkins, J.R. (1987c) Factors inducing inspection of larval honeybees (*Apis mellifera*). (Not yet published).

Free, J.B., Ferguson, A.W. and Simpkins, J.R. (1987d) Queen discrimination by honeybee workers. (Not yet published).

Free, J.B., Ferguson, A.W. and Simpkins, J.R. (1987e) Functions of the alarm pheromone components of the honeybee (*Apis mellifera*). (Not yet published).

Free, J.B., Ferguson, A.W. and Simpkins, J.R. (1987f) Adapting honeybees to synthetic alarm pheromone to reduce aggression. (Not yet published).

von Frisch, K. (1923) Über die 'Sprache' der Bienen, eine Tierpsychologische Untersuchung. *Zoologische Jahrbuch (Physiologie)*, **40**, 1–186.

von Frisch, K. (1967) *The dance language and orientation of bees*. Oxford University Press, London.

von Frisch, K. and Rösch, G.A. (1926) Neue Versuche über die Bedeutung von Duftorgan und Pollenduft für die Verständigung im Bienenvolk. *Zeitschrift für Velgleichende Physiologie*, **4**, 1–21.

Furgala, B. and Boch, R. (1961) Distribution of bees on brood. *Bee World*, **42**, 200–02.

Gary, N.E. (1961a) Queen honeybee attractiveness as related to mandibular gland secretion. *Science*, **133**, 1479–80.

Gary, N.E. (1961b) Antagonistic reactions of worker honeybees to the mandibular gland contents of the queen (*Apis mellifera* L.). *Bee World*, **42**, 14–17.

Gary, N.E. (1961c) Mandibular gland extirpation in living queen and worker honey bees (*Apis mellifera* L.). *Annals of the Entomological Society of America*, **54**, 529–31.

Gary, N.E. (1962) Chemical mating attractants in the queen honeybee. *Science*, **136**, 773–74.

Gary, N.E. (1963) Observations of mating behaviour in the honeybee. *Journal of Apicultural Research*, **2**, 3–13.

Gary, N.E. (1974) Pheromones that affect the behaviour and physiology of honeybees. In *Pheromones* (ed. M.C. Birch) North Holland/Elsevier, Amsterdam, pp. 200–21.

Gary, N.E. and Marston, J. (1971) Mating behaviour of drone honey bees with queen models (*Apis mellifera* L.). *Animal Behaviour*, **19**, 299–304.

Gary, N.E. and Morse, R.A. (1960) A technique for the removal of mandibular glands from living queen honeybees. *Bee World,* **41,** 229–30.

Gary, N.E. and Morse, R.A. (1962) Queen cell construction in honey bee (*Apis mellifera* L.) colonies headed by queens without mandibular glands. *Proceedings of the Royal Entomological Society,* London (A). **37,** 76–8.

Gerig, L. (1972) Ein weiterer Duftstoff zur Anlockung der Drohnen von *Apis mellifica* (L.). *Zeitschrift für angewandte Entomologie,* **70,** 286–89.

Gerig, L. and Gerig, A. (1976) Einsatz ferngesteherter Modellflugzeuge zur Ortung der Drohnenansammlungen von *Apis mellifica* L. *Schweizerische Bienen-Zeitung,* **105,** 379–91.

Getz, W.M., Brückner, D. and Parisian, T.R. (1982) Kin structure and the swarming behaviour of the honey bee *Apis mellifera. Behavioural Ecology and Sociobiology,* **10,** 265–70.

Getz, W.M. and Smith, K.B. (1983) Genetic kin recognition: honey bees discriminate between full and half sisters. *Nature,* **302,** 147–48.

Ghent, R.L. and Gary, N.E. (1962) A chemical alarm release in honey bee stings (*Apis mellifica* L.). *Psyche,* **69,** 1–6.

Greenberg, L. (1979) Genetic component of bee odour in kin recognition. *Science,* **206,** 1095–97.

Groot, A.P. de and Voogd, S. (1954) On the ovary development in queenless worker bees (*Apis mellifera* L.). *Experientia,* **10,** 384–85.

Gunnision, A.F. (1966) *The collection and analysis of the volatiles of bee venom, and solubility characteristics of bee venom solids.* MS Thesis, Cornell University, Ithaca, New York.

Gunnison, A.F. and Morse, R.A. (1968) Source of ether-soluble organics of stings of the honey bee (*Apis mellifera*) (Hymenoptera: Apidae). *Annals of the Entomological Society of America,* **61,** 5–8.

Hass, A. (1946) Neue Beobachtungen zum Problem der Flugbahnen bei Hummelmännchen. *Zeitschrift für Naturforschung,* **11,** 596–600.

Haas, A. (1949a) Regular flight behaviour of the males of *Psithyrus silvestris.* Lep. and several solitary apoidea. *Zeitschrift fur vergleichende Physiologie,* **31,** 671–83.

Haas, A. (1949b) Arttypische Flugbahnen von Hummelmannchen. *Zeitschrift fur vergleichende Physiologie,* **31,** 281–307.

Haas, A. (1952) Die Mandibeldrüse als Duftorgan bei einigen Hymenopteren. *Natürwissenschaften,* **39,** 484.

Hazelhoff, E.H. (1941) De luchtversching van een bijenkast gedurende den zomer. *Maandschrift voor Bijenteelt,* **44,** 16.

Heinrich, B. (1973) Pheromone induced brooding behaviour in *Bombus vosnesenskii* and *Bombus edwardsii* (Hymenoptera: Bombidae). *Journal of the Kansas Entomological Society,* **47,** 396–404.

Hemmling, C., Koeniger, N. and Ruttner, F. (1979) Quantitative Bestimmung der 9-oxodecensäure im Lebenszyklus der Kapbiene (*Apis mellifera capensis* Escholtz). *Apidologie,* **10,** 227–40.

Henrikh, V.G. (1955) Can bees recognise their own queen? (in Russian). *Pchelovodstvo*, **3**, 23–9.

Hess, G. (1942) Partial abstract of experiments concerning laying workers. *Beihefte der Schweizerische Bienenzeitung*, **1**, 33–109.

Hölldobler, B. and Michener, C.D. (1980) Mechanisms of identification and discrimination in social hymenoptera. Dahlem Workshop Report on *'Evolution of Social Behaviour: Hypotheses and Empirical Tests'*. (ed. H. Markl), pp. 35–58.

van Honk, C. and Hogeweg, P. (1981) The ontogeny of the social structure in a captive *Bombus terrestris* colony. *Behavioural Ecology and Sociobiology*, **9**, 111–19.

van Honk, C.G.J., Velthuis, H.H.W. and Röseler, P.F. (1978) A sex pheromone from the mandibular glands in bumblebee queens. *Experientia*, **34**, 838.

van Honk, C.G.J., Velthuis, H.H.W., Röseler, P.F. and Malotaux, M.E. (1980) The mandibular glands of *Bombus terrestris* queens as a source of queen pheromones. *Entomologia Experimentalis et Applicata*, **28**, 191–98.

Huber, F. (1814) *Nouvelles observations sur les abeilles.*2nd Edn. translation 1926, Dadant, Hamilton, Illinois.

Husing, J.O. and Bauer, J. (1968) The extent to which laying workers in bee colonies can be influenced. *Insectes Sociaux*, **15**, 241–44.

Istomina-Tsvetkova, K.P. (1958) Contribution to the study of trophic relations in adult worker bees (in Russian). *17th International Beekeeping Congress*, Apimondia, Bucharest, p. 55–56.

Jacobs, W. (1924) Das Duftorgan von *Apis mellifera* und ähnliche Hhautdrüsenorgane socialer und solitärer Apiden. *Zeitschrift für Morphologie und Ökologie*, **3**, 1–80.

Jay, S.C. (1963) The development of honeybees in their cells. *Journal of Apicultural Research*, **2**, 117–34.

Jay, S.C. (1964) The cocoon of the honey bee *Apis mellifera* L. *The Canadian Entomologist*, **96**, 784–92.

Jay, S.C. (1968) Factors influencing ovary development of worker honeybees under natural conditions. *Canadian Journal of Zoology*, **46**, 345–47.

Jay, S.C. (1970) The effect of various combinations of immature queen and worker bees on the ovary development of worker honeybees in colonies with and without queens. *Canadian Journal of Zoology*, **48**, 169–73.

Jay, S.C. (1972) Ovary development of worker honeybees when separated from worker brood by various methods. *Canadian Journal of Zoology*, **50**, 661–64.

Jay, S.C. (1975) Factors influencing ovary development of worker honeybees of European and African origin. *Canadian Journal of Zoology*, **53**, 1387–90.

Jay, S.C. and Jay, D.H. (1976) The effect of various types of brood comb on the ovary development of worker honeybees. *Canadian Journal of Zoology*, **54**, 1724–26.

Jay, S.C. and Nelson, E.V. (1973) The effects of laying worker honeybees (*Apis mellifera* L.) and their brood on the ovary development of other worker honeybees. *Canadian Journal of Zoology*, **51**, 629–32.

Jaycox, E.R. (1970a) Honey bee queen pheromones and worker foraging behaviour. *Annals of the Entomological Society of America*, **63**, 222–28.

Jaycox, E.R. (1970b) Honey bee foraging behaviour: Responses to queens, larvae, and extracts of larvae. *Annals of the Entomological Society of America*, **63**, 1689–94.

Jaycox, E.R., and Guynn, G. (1973) Effect of anaesthesia of honeybee queens on their attractiveness to workers (in Polish). *Pszczelnicze Zeszyty Naukowe Rok*, **17**, 11–15.

Jaycox, E.R. and Guynn, G. (1974) Comparative development of foraging and brood production in honeybee colonies established on comb and on foundation. *Journal of Apicultural Research*, **13**, 229–33.

Juska, A. (1978) Temporal decline in attractiveness of honeybee queen tracks. *Nature*, **276**, 261.

Juska, A., Seeley, T.D. and Velthuis, H.H.W. (1981) How honeybee attendants become ordinary workers. *Journal of Insect Physiology*, **27**, 515–19.

Kaissling, K.E. and Renner, M. (1968) Antennale Rezeptoren für Queen Substance und Sterzelduft bei der Honigbiene. *Zeitschrift fur Vergleichende Physiologie*, **59**, 357–61.

Kalmus,H. and Ribbands, C.R. (1952) The origin of the odours by which honeybees distinguish their companions. *Proceedings of the Royal Society, B.*, **140**, 50–9.

Kashkovskii, V.G. (1957) The influence of the removal of the queen on the productivity of bee colonies (in Russian). *Pchelovodstvo*, **34**, 36–8.

Keeping, M.G., Crewe, R.M. and Field, B.I. (1982) Mandibular gland secretions of the old world stingless bee, *Trigona gribodoi* Magretti: isolation, identification and compositional changes with age. *Journal of Apicultural research*, **21**, 65–73.

Kerr, W.E. (1969) Some aspects of the evolution of social bees. *Evolutionary Biology*, **3**, 119–75.

Kerr, W.E. (1973) Sun compass orientation in the stingless bees *Trigona (Trigona) spinipes* (Fabricius, 1783) (Apidae). *Anais da Academia Brasileira de Ciencias*, **45**, 301–08.

Kerr, W.E., Ferreira, A. and Simoes de Mattos, N. (1963) Communication among stingless bees – additional data (Hymenoptera: Apidae). *Journal of the New York Entomological Society*, **71**, 80–90.

Kerr, W.E., Blum, M.S., Pisani, J.F. and Short, A.C. (1974) Correlation between amounts of 2-heptanone and iso-amyl acetate in honeybees and their aggressive behaviour. *Journal of Apicultural Research*, **13**, 173–76.

Kerr, W.E., Blum, M.S. and Fales, H.M. (1981) Communication of food source between workers of *Trigona (Trigona) spinipes*. *Revista Brasileira de Biologia*, **41**, 619–23.

Kigatiira, K.I. (1984) *The aspects of the ecology of the African honeybee*. Ph.D. Thesis. University of Cambridge.

Kigatiira, K.I. Beament, J.W.L., Free, J.B. and Pickett, J.A. (1986) Using synthetic pheromone lures to attract honeybee colonies in Kenya. *Journal of Apicultural Research* **25**, 85–86.

Koeniger, N. (1970) Factors determining the laying of drone and worker eggs by the queen honeybee. *Bee World*, **51**, 166–69.

Koeniger, N. (1978) Das Warmen der Brut bei der Honigbiene. *Apidologie*, **9**, 305–20.

Koeniger, N. (1984) Brood care and recognition of pupae in the honeybee (*Apis mellifera*) and the hornet (*Vespa crabro*). Symposium on 'Insect Communication'. *Royal Entomological Society of London*. pp. 267–82.

Koeniger, N. and Koeniger, G. (1980) Observations and experiments on migration and dance communication of *Apis dorsata* in Sri Lanka. *Journal of Apicultural Research*, **19**, 21–34.

Koeniger, N. and Veith, H.J. (1983) Glyceryl-1,2-dioleate-3-palmitate, a brood pheromone of the honeybee (*Apis mellifera* L.). *Experientia*, **39**, 1051–52.

Koeniger, N. and Veith, H.J. (1984) Spezifität eines Brutpheromons und Bruterkennung bei der Honigbiene (*Apis mellifera* L.). *Apidologie*, **15**, 205–10.

Koeniger, N. and Wijayagunasekera, H.N.P. (1976) Time of drone flight in three Asiatic honeybee species. *Journal of Apicultural Research*, **15**, 67–71.

Koeniger, G., Koeniger, N. and Fabritius, M. (1979a) Some detailed observations of mating in the honeybee. *Bee World*, **60**, 53–7.

Koeniger, N., Weiss, J. and Maschwitz, U. (1979b) Alarm pheromones of the sting in the genus *Apis*. *Journal of Insect Physiology*, **25**, 467–76.

Koptev, V.S. (1957) Laying workers and swarming (in Russian). *Pchelovodstvo*, **34**, 31–2.

Kropáčova, S. and Haslbachová, H. (1970) The development of ovaries in worker honeybees in queenright colonies examined before and after swarming. *Journal of Apicultural Research*, **9**, 65–70.

Kropáčova, S. and Haslbachová, H. (1971) The influence of queenlessness and of unsealed brood on the development of ovaries in worker honeybees. *Journal of Apicultural Research*, **10**, 57–61.

Kruger, E. (1951) Uber Bahnfluge der Mannchen der Gattungen *Bombus* und *Psithyrus*. *Zeitschrift für Tierpsychologie*, **8**, 61–75.

Kubišová, S. and Haslbachová, H. (1978) Effects of larval extracts on the development of ovaries in caged worker honeybees. *Acta Entomologica Bohemoslovaca*, **75**, 9–14.

Kubišová, S., Haslbachová, H. and Vrkroč, J. (1982) Effects of fractions of larval extracts on the development of ovaries in caged worker honey bees. *Acta Entomologica Bohemoslovaca*, **79**, 334–40.

Kullenberg, B. (1956) Field experiments with chemical sexual attractants on aculeate hymenoptera males I. *Zoologiska Bidrag. Fran Uppsala*, **31**, 254–352.

Kullenberg, B. (1973) Field experiments with chemical sexual attractants on aculeate hymenoptera males II. *Zoon Supplement*, **1**, 31–42.

Kullenberg, B., Bergström, G. and Stallberg-Stenhagen, S. (1970) Volatile components of the cephalic marking secretion of male bumblebees. *Acta Chemica Scandinavica*, **24**, 1481–83.

Kullenberg, B., Bergström, G., Bringer, B., Carlberg, B. and Cederberg, B. (1973) Observations on scent marking by *Bombus* Latr. and *Psithyrus* Lep. males (Hym. Apidae) and localisation of site of production of the secretion. *Zoon Supplement*, **1**, 23–30.

Lavie, P. and Pain, J. (1959) Biologie – relation entre la substance attractive, le facteur antibiotique et le développement ovarien chez la reine d'abeille *Apis mellifica*. *Comptes Rendus des Séanes de L'Academie de Science*, **248**, 3753–55.

Lecomte, J. (1956) Über die Bildung von 'Strassen' durch Sammelbienen, deren Stock um 180° gedreht wurde. *Zeitschrift für Bienenforschung*, **3**, 128–33.

Lecomte, J. (1957) Sur le marquage olfactif des sources de nourriture par les abeilles butineuses. *Compte Rendu de L'Académie de Science de Paris*, **245**, 2385–87.

Lensky, Y. and Slabezki, Y. (1981) The inhibiting effect of the queen bee (*Apis mellifera* L.) foot-print pheromone on the construction of swarming queen cups. *Journal of Insect Physiology*, **27**, 313–23.

Lesher, C. and Morse, R.A. (1983) Baits to improve bait hive attractiveness to honey bees. *American Bee Journal*, **123**, 193–94.

Levchenko, I.A. and Shalimov, I.I. (1975) Exposure of Nasonov gland in honey bees when walking towards the food source. *25th International Apicultural Congress, Apimondia*, p. 305.

Lindauer, M. (1948) Über die Einwirkung von Duft-und Geschmacksstoffen sowie andere Faktoren auf die Tänze der Bienen. *Zeitschrift für Vergleichende, Physiologie* **31**, 348–412.

Lindauer, M. (1956) Über die Verstandingung bei indischen Bienen. *Zeitschrift fur Vergleichende Physiologie*, **38**, 521–57.

Lindauer, M. (1957) Communication in swarm-bees searching for a new home. *Nature*, **179**, 63–6.

Lindauer, M. and Kerr, W.E. (1958) Die Gegenseitige Verständigung bei den Stachellosen Bienen. *Zeitschrift für Vergleichende Physiologie*, **41**, 405–34.

Lundie, A.E. (1954) Laying worker bees produce worker bees. *South African Bee Journal*, **29**, 10–11.

Mackenson, O. (1947) Effect of carbon dioxide on initial oviposition of artificially inseminated and virgin queens. *Journal of Economic Entomology*, **40**, 344–49.

McGregor, S.E. (1976) Insect pollination of cultivated crop plants. *Agricultural Handbook, United States Department of Agriculture*, No. 496, 411 pp.

McIndoo, N.E. (1914) The scent-producing organ of the honey bee. *Proceedings of the Society of Natural Science and Philosophy*, **66**, 542–55.

Maschwitz, U. (1964) Gefahrenalarmstoffe und Gefahrenalarmierung bei sozialen Hymenopteren. *Zeitschrift für Vergleichende Physiologie*, **47**, 596–655.

Mathis, M. (1947) Research notes. *Bee World*, **28**, 81.

Maurizio, A. (1954) Pollenernährung und Lebensvorgänge bei der Honigbiene. *Landwirtschaft Jahrbuch Schweizerei*, **68**, 115–82.

Michener, C.D. (1972) *Final report, committee on the African honeybee*. National Academy of Science, Washington DC.

Michener, C.D., Brothers, D.J. and Kamm, D.R. (1971) Interactions in colonies of primitively social bees: artificial colonies of *Lasio glossum zephyrum*. *Proceedings of the National Academy of Science*, **68**, 1241–45.

Milne, C.P. (1983a) Honey bee (Hymenoptera: Apidae) hygienic behaviour and resistance to chalkbrood. *Annals of the Entomological Society of America*, **76**, 384–87.

Milne, C.P. (1983b) Laboratory measurement of honeybee brood disease resistance. 2. uncapping of freeze-killed and live brood by newly emerged workers in cages. *Journal of Apicultural Research*, **22**, 115–18.

Milojević, B.D., and Filipović-Moskovljević, V. (1958) De l'effet de groupe chez les abeilles domestiques. *17th International Beekeeping Congress*, p. 82–3.

Milojević, B.D., and Filipović-Moskovljević, V. (1959) Gruppeneffekt bei Honigbienen. II. Eierstockentwicklung bei Arbeitsbienen im Kleinvolk. *Bulletin de l'Académie Serbe des Sciences. Classe des Sciences Mathematiques et Naturelles*, **25**, 131–138.

Milojević, B.D., Filipović-Moskovljević, V. and Djakovic, D. (1963) Dependence of queen's influence in honeybee society upon the phase of queen's life. *Bulletin de L'Académie Serbe des Sciences. Classe des Sciences Mathematiques et Naturelles*, **32**, 45–9.

Moeller, F.E. (1961) The relationship between colony populations and honey production as affected by honey bee stock lines. *United States Department of Agriculture Production and Research Report*, **55**, 1–20.

Moeller, F.E. (1976) Two-queen system of honey-bee colony management. United States Department of Agriculture Production Research Report, **161**, 1–11.

Moeller, F.E. and Harp, E.R. (1965) The two-queen system simplified. *Gleanings of Bee Culture*, **93**, 679–82, 698.

Morse, R.A. (1963) Swarm orientation in honeybees. *Science*, **141**, 357–58.

Morse, R.A. (1966) Honey bee colony defense at low temperatures. *Journal of Economic Entomology*, **59**, 1091–93.

Morse, R.A. (1972) Honey bee alarm pheromone; another function. *Annals of the Entomological Society of America*, **65**, 1430.

Morse, R.A., and Boch, R. (1971) Pheromone concert in swarming honeybees (Hymenoptera: Apidae). *Annals of the Entomological Society of America*, **64**, 1414–17.

Morse, R.A. and Gary, N.E., (1963) Further studies of the responses of honey bee (*Apis mellifera* L.) colonies to queens with extirpated mandibular glands. *Annals of the Entomological Society of America*, **56**, 372–74.

Morse, R.A. and Laigo, F.M. (1969) *Apis dorsata* in the Philippines. *Monograph from the Philippine Association of Entomology*, **1**, 1–96.

Morse, R.A. and McDonald, J.L. (1965) The treatment of capped queen cells by honeybees. *Journal of Apicultural Research*, **4**, 31–4.

Morse, R.A., Gary, N.E. and Johansson, T.S.K. (1962) Mating of virgin queen honey bees (*Apis mellifera* L.) following mandibular gland extirpation. *Nature*, **194**, 605.

Morse, R.A., Macdonald, J.E. and Zmarlicki, C. (1963) The effect of queen cell construction in the honey bee (*Apis mellifera* L.) colony. *The Florida Entomologist*, **46**, 219–21.

Morse, R.A., Shearer, D.A., Boch, R. and Benton, A.W. (1967) Observations on alarm substances in the genus *Apis*. *Journal of Apicultural Research*, **6**, 113–18.

Müssbichler, A. (1952) Die Bedeutung äusserer Einflüsse und der Corpora Allata bei der Afterweisel-entstehung von *Apis mellifera*. *Zeitschrift für Vergleichende Physiologie*, **34**, 207–21.

Nedel, J.O. (1960) Morphologie und Physiologie der Mandibeldrüse Einger Bienen-Arten (Apidae). *Zeitschrift für Morphologie und Ökologie*, **49**, 139–83.

Newman, H.W. (1851) Habits of the Bombinatrices. *Proceedings of the Entomological Society, London*, **1**, 86–92.

Newton, D.C. and Michl, D.J. (1974) Cannibalism as an indication of pollen insufficiency in honeybees: Ingestion or recapping of manually exposed pupae. *Journal of Apicultural Research*, **13**, 235–41.

Newton, D.C. (1968) Behavioural response of honeybees to colony disturbance by smoke. I. Engorging behaviour. *Journal of Apicultural Research*, **7**, 3–9.

Newton, D.C. (1971) When bees are smoked. *American Bee Journal*, **111**, 180–82.

Nixon, H.L. and Ribbands, C.R. (1951) Food transmission within the honeybee community. *Proceedings of the Royal Society*, **140**, 43–50.

Núñez, J.A. (1967) Sammelbienen markieren versiegte Futterquellen durch Duft. *Naturwissenschaften*, **54**, 322–23.

Pain, J. (1954a) Sur l'ectohormone des reines d'Abeilles. *Comptes Rendus de L'Académie de Science* (Paris), **239**, 1869–70.

Pain, J. (1954b) La 'substance de fécondité' dans le développement des ovaries des ouvrières d'abeilles (*Apis mellifica* L.) *Insectes Sociaux*, **1**, 59–70.

Pain, J. (1955) Dosage biologique et spectrographie de l'ectohormone des reines d'abeilles. *Comptes Rendus des Séances L'Académie de Science*, Paris, **240**, 670–72.

Pain, J. (1956) Mesure du pouvoir inhibiteur et de l'attractivité de l'ectohormone des reines d'Abeilles. Différences individuelles. *Comptes Rendus des Seances de L'Académie de Science*, **242**, 1080–82.

Pain, J. (1961a) Sur la phéromone des reines d'Abeilles et ses effets physiologiques. *Annales de L'Abeille*, **4**, 73–152.

Pain, J. (1961b) L'importance du pouvoir d'attraction des reines d'abeilles pour le prélèvement de la phéromone. *Symposium di Genetica Biologia*, Italia, **10**, 189–97.

Pain, J. (1961c) Absence du pouvoir d'inhibition de la phéromone. I. Sur le développement ovarien des jeunes ouvrières d'abeilles. *Comptes Rendus des Séances de L'Académie de Science*, Paris, **252**, 2316–17.

Pain, J. and Barbier, M. (1981) The pheromone of the queen honeybee. *Naturwissenschaften*, **67**, 429.

Pain, J., Barbier, M. and Roger, B. (1967) Dosages individuels des acides céto-9-décène-2-oïque et hydroxy-10-décène-2-oïque dans les têtes des reines et des ouvrières d'abeilles. *Annales de L'Abeille*, **10**, 45–52.

Pain, J. and Roger, B. (1967) Variation de la teneur en acide céto-9-décène-2-oïque en fonction de l'âge cez les reines vierges d'abeille (*Apis mellifica ligustica* S.). *Comptes Rendus des Seánces de L'Académie de Science*, Paris, **283**, 797–99.

Pain, J. and Roger, B. (1978). Rhythme circadien des acides céto-9-décène-2-oïque, phéromone de la reine, et hydroxy-10 décène-2-oïque des ouvrières d'abeilles, *Apis mellifica ligustica* S. *Apidologie*, **9**, 263–72.

Pain, J. and Ruttner, F. (1963) Les extraits de glandes mandibulaires des reines d'abeilles attirent les males lors du vol nuptial. *Comptes Rendus Hebdomadaires des Seances de L'Académie de Science*, Paris, **256**, 512.

Pain, J., Roger, B. and Theurkauff, J. (1972) Sur l'existence d'un cycle annuel de la production de phéromone (acide céto-9-décène-2-oïque) chez les reines d'abeilles (*Apis mellifica ligustica* Spinola). *Comptes Rendus des Séances de L'Académie de Science*, Paris, **275**, 2399–2402.

Park, O.W. (1949) The honeybee colony – life history. In *The Hive and the Honeybee*, (ed. R.A. Groot), Dadant, Hamilton, Illinois.

Perepelova, L.I. (1926) Biology of laying workers (in Russian). *Opytnaya Pasieka*, (12), 8–10.

Perepelova, L.I. (1928) Biology of laying workers. I. Relationship of the age of bees to the development of workers. (in Russia). *Opytnaya Pasieka*, (1), 6–10.

Pflumm, W. (1969) Beziehungen zwischen Putzvedrhalten und Sammelbereitschaft bei der Honigbiene. *Zeitschrift für vergleichende Physiologie*, **64**, 1–36.

Pflumm, W. and Wilhelm, K. (1982) Olfactory feedback in the scent-marking behaviour of foraging honeybees at the food source? *Physiological Entomology*, **7**, 203–07.

Pflumm, W., Peschke, C., Wilhelm, K. and Cruse, H. (1978) Einfluss der – in einem Flugraum kontrollierten – Trachtverhaltnisse auf das Duftmarkieren und die Abflugmagenfullung der Sammelbiene. *Apidologie*, **9**, 349–62.

Pham, M.H., Roger, B. and Pain, J. (1983) Attractiveness of a queen pheromonal extract to worker bees of varying ages (*Apis mellifica ligustica* S.). *Apidologie*, **14**, 152–55.

Pickett, J.A., Williams, I.H., Martin, A.P. and Smith, M.C. (1980) Nasonov pheromone of the honeybee, *Apis mellifera* L. (Hymenoptera: Apidae) Part 1. Chemical characterisation. *Journal of Chemical Ecology*, **6**, 425–34.

Pickett, J.A., Williams, I.H., Smith, M.C. and Martin, A.P. (1981) The Nasonov pheromone of the honey bee, *Apis mellifera* L. (Hymenoptera: Apidae) Part III. Regulation of pheromone composition and production. *Journal of Chemical Ecology*, **7**, 543–54.

Pickett, J.A., Williams, I.H. and Martin, A.P. (1982) (Z)-11-eicosen-1-ol, an important new pheromonal component from the sting of the honey bee, *Apis mellifera* L. (Hymenoptera: Apidae). *Journal of Chemical Ecology*, **8**, 163–75.

Pickett, J.A., Dawson, G.W., Griffiths, D.C., Liu, X., Macaulay, E.D.M. and Woodcock, C.M. (1985) Propheromones: an approach to the slow release of pheromones. *Pesticide Science*, **15**, 261–64.

Pomeroy, N. (1981) *Reproductive dominance interactions and colony development in bumblebees (Bombus latreille;* Hymenoptera: Apidae). Ph.D. Thesis, University of Toronto.

Rajashekharappa, B.J. (1979) Queen recognition and rearing by honeybee (*Apis cerana* Fabricius) colonies. *Journal of Apicultural Research*, **18**, 173–78.

Ramli, J., Sulaiman, M.R. and Abdullah, N. (1984) The effects of synthetic pheromones on clustering activity of *Apis dorsata* (Giant honeybee). *1st Regional Symposium for Biological Control*, Serdang. Abstract No. **39**, p. 67.

Renner, M. (1955) Neue Untersuchungen über die Physiologische Wirkung des Duftorganes der Honigbiene. *Naturwissenschaften*, **42**, 589.

Renner, M. (1960) Das Duftorgan der Honigbiene und die physiologische

Bedeutung inres Lockstoffes. *Zeitschrift für vergleichende Physiologie*, **43**, 411–68.

Renner, M. and Baumann, M. (1964) Über Komplexe von subepidermalen Drüsenzellen (Duftdrüsen) der Bienenkönigin. *Naturwissenschaften*, **51**, 68–9.

Renner, M. and Vierling, G. (1977) Die Rolle des Taschendrüsenpheromons beim Hochzeitsflug der Bienenkönigin. *Behavioural Ecology and Sociobiology*, **2**, 329–38.

Ribbands, C.R. (1953) *The behaviour and social life of honeybees*, Bee Research Association, London.

Ribbands, C.R. (1954) Communication between honeybees. 1. The response of crop-attached bees to the scent of their crop. *Proceedings of the Royal Entomological Society*, London. **29**, 141–44.

Ribbands, C.R. (1955) The scent perception of the honeybee. *Proceedings of the Royal Society B*, **143**, 367–79.

Ribbands, C.R. and Speirs, N. (1953) The adaptability of the homecoming honeybee. *British Journal of Animal Behaviour*, **1**, 59–66.

Rinderer, T.E. (1981) Volatiles from empty comb increase hoarding by the honeybee. *Animal Behaviour*, **29**, 1275–76.

Rinderer, T.E. (1982) Maximal stimulation by comb of honey bee (*Apis mellifera*) hoarding behaviour. *Annals of the Entomological Society of America*, **75**, 311–12.

Rinderer, T.E. and Baxter, J.R. (1978) Effect of empty comb on hoarding behaviour and honey production of the honey bee. *Journal of Economic Entomology*, **71**, 757–59.

Rinderer, T.E. and Baxter, J.R. (1979) Honey bee hoarding behaviour: effects of previous stimulation by empty comb. *Animal Behaviour*, **27**, 426–28.

Rinderer, T.E. and Baxter, J.R. (1980) Amount of empty comb, comb colour and honey production. *Americal Bee Journal*, **120**, 641–42.

Rinderer, T.E. and Hagstad, W.A. (1984) The effect of empty comb on the proportion of foraging honeybees collecting nectar. *Journal of Apicultural Research*, **23**, 80–81.

Rinderer, T.E., Dolten, A.B., Harbo, J.R. and Collins, A.M. (1982) Hoarding behaviour of European and Africanised honeybees (Hymenoptera: Apidae). *Journal of Economic Entomology*, **75**, 714–15.

Rinderer, T.E., Hellmich, II, R.L., Danka, R.G. and Collins, A.M. (1985) Male reproductive parasitism: a factor in the Africanisation of European honeybee populations. *Science*, **228**, 1119–21.

Ritter, W. (1981) *Varroa* disease of the honeybee *Apis mellifera*. *Bee World*, **62**, 141–53.

Röseler, P.F. (1970) Unterschiede in der Kastendetermination zwischen den Hummelarten *Bombus hypnorum* und *Bombus terrestris*. *Zeitschrift für*

Naturforschung, **25**, 543–48.

Röseler, P.F. (1974) Vergleichende Untersuchungen zur Oögenese bei weiselrichtigen und weisellosen Arbeiterinnen der Hummelart *Bombus terrestris*. *Insectes Sociaux*, **21**, 249–74.

Röseler, P.F. (1977) Juvenile hormone control of oögenesis in bumblebee workers, *Bombus terrestris*. *Journal of Insect Physiology*, **23**, 985–92.

Röseler, P.F. and Röseler, I. (1978) Studies on the regulation of the juvenile hormones titre in bumblebee workers, *Bombus terrestris*. *Journal of Insect Physiology*, **24**, 707–13.

Röseler, P.F., Röseler, I. and van Honk, C.G.J. (1981) Evidence for corpora allata activity in workers of *Bombus terrestris* by a pheromone from the queen's mandibular glands. *Experientia*, **37**, 348–51.

Rothenbuhler, W.C. (1964) Behaviour genetics of nest cleaning in honey bees. I. Responses of four inbred lines to disease-killed brood. *Animal Behaviour*, **12**, 578–83.

Ruttner, F. (1956) The mating of the honeybee. *Bee World*, **37**, 315.

Ruttner, F. (1966) The life and flight activity of drones. *Bee World*, **47**, 93–100.

Ruttner, F. and Kaissling, K.E. (1968) Über die interspezifische Wirkung des Sexuallockstoffes von *Apis mellifica* und *Apis cerana*. *Zeitschrift für vergleichende Physiologie*, **59**, 362–70.

Ruttner, F. and Maul, V. (1983) Experimental analysis of reproductive interspecies isolation of *Apis mellifera* L. and *Apis cerana* Fabr. *Apidologie*, **14**, 309–27.

Ruttner, F. and Ruttner, H. (1965) Untersuchungen über die Flugaktivität und das Paarungsverhalten der Drohnen. 2. Beobachtungen an Drohnen-sammelplätzen. *Zeitschrift für Bienenforschung*, **8**, 1–9.

Ruttner, F. and Ruttner, H. (1966) Untersuchungen über die Flugaktitität und das Paarungsverhalten der Dorhnen. *Zeitschrift für Bienenforschung*, **8**, 332–54.

Ruttner, F. and Ruttner, H. (1968) Untersuchungen über die Flugaktivität und das Paarungsverhalten der Drohnen. 4. Zur Fernorientierung und Orisstetigkeit der Drohnen auf ihren Paarungsflugen. *Zeitschrift für Bienenforschung*, **9**, 259–65.

Ruttner, F., Woyke, J. and Koeniger, N. (1972) Reproduction in *Apis cerana*. 1. Mating behaviour. *Journal of Apicultural Research*, **11**, 141–46.

Ruttner, F., Koeniger, N. and Veith, H.J. (1976) Queen substance bei eierlegenden Arbeiterinnen der Honigbiene (*Apis mellifica* L.). *Naturwissenschaften*, **9**, 434.

Saiovici, M. (1983) 9-oxodecenoic acid and dominance in honeybees. *Journal of Apicultural Research*, **22**, 27–32.

Sakagami, S.F. (1954) Occurrence of an aggressive behaviour in queenless hives with considerations on the social organisation of honeybees. *Insectes Sociaux*, **1**, 331–43.

Sakagami, S.F. (1958) The false queen: Fourth adjustive response in dequeened honeybee colonies. *Behaviour*, **13**, 280–96.

Sakagami, S.F. (1959) Arbeitsteilung in einem Weisellosen Bienenvölkchen. *Zeitschrift für Bienenforschung*, **4**, 186–93.

Salmon, A.W. (1938) Virgin queens. *British Bee Journal*, **66**, 62.

Sannasi, A., Ratulu, G.S. and Sundara, G. (1971) 9-oxo-trans-2-decenoic acid in the Indian honeybees. *Life Science*, **10**, 195–201.

Schmidt, J.O. and Thoenes, S.C. (1987) A trap for potential survey and control of Africanized honeybee (*Hymenoptera: Apidae, Apis mellifera scutellata*) swarms. Bulletin of the Entomological Society of America (in press).

Schremmer, F. (1972) Beobachtungen zum Paarungsverhalten der Männchen von *Bombus confusus* Schenk. *Zeitschrift für Tierpsychologie*, **31**, 503–12.

Seeley, T.D. (1979) Queen substance dispersal by messenger workers in honeybee colonies. *Behavioural Ecology and Sociobiology*, **5**, 391–415.

Seeley, T.D. and Fell, R.D. (1981) Queen substance production in honey bee (*Apis mellifera*) colonies preparing to swarm (Hymenoptera: Apidae). *Journal of the Kansas Entomological Society*, **54**, 192–96.

Shaposhnikova, N.G. and Gavrilov, B.N. (1973) Perception by honeybee workers (*Apis mellifera caucasia*) of synthetic queen substance (9-oxodec-trans-2-enoic acid) (in Russian). *Zoologicheskii Zhurnal*, **52**, 291–93.

Shaposhnikova, N.G. and Gavrilov, B.N. (1974) The effect of a synthetic pheromone on bees (in Russian). *Pchelovodstvo*, **94**, 23.

Shaposhnikova, G. and Gavrilov, B.N. (1975) Increasing the attractiveness of queens (to workers) and their introduction into colonies (in Russian). *Insect Chemoreception No. 2*. (ed. A.V. Skirkyavichyus) 181–84.

Shearer, D.A. and Boch, R. (1965) 2-heptanone in the mandibular gland secretion of the honey bee. *Nature*, **206**, 530.

Shearer, D.A. and Boch, R. (1966) Citral in the Nassonoff pheromone of the honey bee. *Journal of Insect Physiology*, **12**, 1513–21.

Shearer, D.A., Boch, R., Morse R.A. and Laigo, F.M. (1970) Occurrence of 9-oxodec-trans-2-enoic acid in queens of *Apis dorsata, Apis cerana* and *Apis mellifera*. *Journal of Insect Physiology*, **16**, 1437–41.

Simpson, J. (1956) *Annual Report 1956*. Bee Department, Rothamsted Experimental Station, p. 161.

Simpson, J. (1957) The incidence of swarming among colonies of honeybees in England. *Journal of Agricultural Science*, **49**, 387–93.

Simpson, J. (1959) Variation in the incidence of swarming among colonies of *Apis mellifera* throughout the summer. *Insectes Sociaux*, **6**, 85–99.

Simpson, J. (1960) Induction of queen rearing in honeybee colonies by amputation of their queens' front legs. *Bee World*, **4**, 286–87.

Simpson, J. (1963) Queen perception by honey bee swarms. *Nature*, **99**, 94–5.

Simpson, J. (1966) Repellency of the mandibular gland scent of worker honey bees. *Nature*, **209**, 531–32.

Simpson, J. (1979) The existence and physical properties of pheromones by which worker honeybees recognise queens. *Journal of Apicultural Research*, **18**, 232–49.

Sladen, F.W.L. (1901) A scent organ in the bee. *British Bee Journal*, **29**, 142, 143, 151–53.

Sladen, F.W.L. (1902) A scent producing organ in the abdomen of the worker of *Apis mellifera*. *Entomologists Monthly Magazine*, **38**, 208–211.

Sladen, F.W.L. (1905) *Queen rearing in England*. Houlson, London.

Sladen, F.W.L. (1912) *The Humble-bee. Its Life History and How to Domesticate it*. Macmillan, London.

Smith, M.V. (1974) Relationship of age to brood-rearing activities of worker honeybees *Apis mellifera* L. *Proceedings of the Society of Ontario 1974*, **105**, 128–32.

Smith, B.H. and Roubik, D.W. (1983) Mandibular glands of stingless bees (Hymenoptera: Apidae): Chemical analysis of their contents and biological function in two species of *Melipona*. *Journal of Chemical Ecology*, **9**, 1465–72.

Snodgrass, R.E. (1956) *Anatomy of the Honeybee* Constable, London.

Stein, G. (1963) Über den Sexuallockstoff von Hummelmännchen. *Naturwissenschaften*, **50**, 305.

Stort, A.C.G. (1972) Metodologia para o estudo da genetica da aggressividade de *Apis mellifera*. *Proceedings of the 1st Congresso Brasileiro de Apicultura (Florianopolis)* p. 36–50.

Stort, A.C.G. (1974) Genetic study of aggressiveness of two subspecies of *Apis mellifera* in Brazil. 1. Some tests to reduce aggressiveness. *Journal of Apicultural Research*, **13**, 33–8.

Strang, G.E. (1970) A study of honey bee drone attraction in the mating response. *Journal of Economic Entomology*, **63**, 641–45.

Svensson, Bo. G. (1979) *Pyrobombus lapponicus* Auct., in Europe recognised as two species: *P. lapponicus* (Fabricius, 1793) and *P. monticola* (Smith, 1849) (Hymenoptera: Apoidea: Bombinae). *Entomologica Scandanavica*, **10**, 275–96.

Svensson, B.G. and Bergström, G. (1977) Volatile marking secretions from the labial gland of north European *Pyrobombus D.T.* males (Hymenoptera: Apidae). *Insectes Sociaux*, **24**, 213–24.

Svensson, B.G. and Bergström, G. (1979) Marking pheromones of *Alpinobombus* males. *Journal of Chemical Ecology*, **5**, 603–15.

Szabo, T.I. and Smith, M.V. (1973) Behavioural studies on queen introduction in honey bees (*Apis mellifera* L.) V. Behavioural relationship between pairs of queens without worker attendants. *Proceedings of the Entomological Society of Ontario*, **103**, 87–96.

Taber, S. and Owens, C.D. (1970) Colony founding and initial nest design of honey bees *Apis mellifera* L. *Animal Behaviour*, **18**, 625–32.

Taranov,G.F. (1959) The production of wax in the honey bee colony. *Bee World*, **40**, 113–21.

Todd, F.E. and Reed, C.B. (1970) Brood measurement as a valid index to the value of honeybees as pollinators. *Journal of Economic Entomology*, **63**, 148–49.

Tribe, G.D. (1982) Drone mating assemblies. *South African Bee Journal*, **54**, 99–100, 103–12.

Vareschi, E. (1971) Duftunterscheidung bei der Honigbiene – einzelzell – ableitungen und verhaltens Reaktionen. *Zeitschrift für vergleichende Physiologie*, **75**, 143–73.

Vecchi, M.A. (1960) La glandola odoripara dell *Apis mellifica* L. *Bollettino Dell'Istituto di Entomologia Della Università di Bologna*, **24**, 53–66.

Veith, J., Weiss, J. and Koeniger, N. (1978) A new alarm pheromone (2-decen-1-yl-acetate) isolated from the stings of *Apis dorsata* and *Apis florea* (Hymenoptera: Apidae). *Experientia*, **34**, 423.

Velthuis, H.H.W. (1967) On abdominal pheromones; the queen honeybee. *Proceedings of the 21st International Beekeeping Congress, College Park, Maryland, USA*, p. 472.

Velthuis, H.H.W. (1970a) Queen substance from the abdomen of the honeybee queen. *Zeitschrift für vergleichende Physiologie*, **70**, 210–22.

Velthuis, H.H.W. (1970b) Ovarian development in *Apis mellifera* worker bees. *Entomologia Experimentalis et Applicata*, **13**, 377–94.

Velthuis, H.H.W. (1972) Observations on the transmission of queen substances in the honeybee colony by the attendants of the queen. *Behaviour*, **41**, 105–29.

Velthuis, H.H.W. (1976) Egg laying, aggression and dominance in bees. *Proceedings of the 15th International Congress of Entomology, Washington, 1976*, pp. 436–49.

Velthuis, H.H.W. (1985) The honeybee queen and the social organization of her colony. *Fortschritte de Zoologie*, **31**, 343–357.

Velthuis, H.H.W. and van Es, J. (1964) Some functional aspects of the mandibular glands of the queen honeybee. *Journal of Apicultural Research*, **3**, 11–16.

Velthius, H.H.W., Verheijen, F.J. and Gottenbos, A.J. (1965) Laying worker honey bee: similarities to the queen. *Nature*, **207**, 1314.

Velthuis, H.H.W., Clement, J., Morse, R.A. and Laigo, F.M. (1971) The ovaries of *Apis dorsata* workers and queens from the Philippines. *Journal of Apicultural Research*, **10**, 65–6.

Velthuis, H.H.W., Ruttner, F. and Crewe, R.M. (1981) Differentiation in physiology and behaviour during the development of laying worker honeybees. In *Social Insects, and Evolutionary Approach to Caste and Reproduction* (ed. W. Engels), Heidelberg, New York.

Verheijen-Voogd, C. (1959) How worker bees perceive the presence of their queens. *Zeitschrift für vergleichende Physiologie*, **41**, 527–82.

Vierling, G. and Renner, M. (1977) Die Bedeutung des Sekretes der Tergittaschendrüsen für die Attraktivität der Bienenkönigin gegenüber junger Arbeiterinnin. *Behavioural Ecology and Sociobiology*, **2**, 185–200.

Voodg, S. (1955) Inhibition of ovary development in worker bees by extraction fluid of the queen. *Experientia*, **11**, 181–86.

Voogd, S. (1956) The influence of a queen on the ovary development in worker bees. *Experientia*, **12**, 199–201.

Waller, G.D. (1970) Attracting honeybees to alfalfa with citral, geraniol and anise. *Journal of Apicultural Research*, **9**, 9–12.

Waller, G.D. (1973) The effect of citral and geraniol conditioning on the searching activity of honeybee recruits. *Journal of Apicultural Research*, **12**, 53–7.

Weaver, N., Weaver, E.C. and Law, J. (1964) The attractiveness of citral to foraging honeybees. *Progress Report, Texas A & M University/Texas Agricultural Experimental Station*. p. 1–6.

Wedmore, E.B. (1932) A manual of beekeeping for English-speaking beekeepers. Edward Arnold, London.

Wenner, A.M., Wells, P.H. and Johnson, D.L. (1969) Honeybee recruitment to food sources: olfaction or language? *Science*, **164**, 84–6.

Wilhelm, K.T. and Pflumm, W.W. (1983) Über den Einfluss Blumenhafter Düfte auf das Duftmarkieren der Sammelbiene. *Apidologie*, **14**, 183–90.

Williams, I.H. and Free, J.B. (1975) Effect of environmental conditions during the larval period on the tendency of worker honeybees to develop their ovaries. *Journal of Entomology*, **49**, 179–82.

Williams, I.H., Pickett, J.A. and Martin, A.P. (1981) The Nasonov pheromone of the honeybee *Apis mellifera* L. (Hymenoptera: Apidae) part II. Bioassay of the components using foragers. *Journal of Chemical Ecology*, **7**, 225–37.

Williams, I.H., Pickett, J.A. and Martin, A.P. (1982) Nasonov pheromone of the honey bee *Apis mellifera* L. (Hymenoptera: Apidae) IV Comparative electroantennogram responses. *Journal of Chemical Ecology*, **8**, 567–74.

Woyke, J. (1977) Cannibalism and brood-rearing efficiency in the honeybee. *Journal of Apicultural Research*, **16**, 84–94.

Woyke, J. (1980) Evidence and action of cannibalism substance in *Apis cerana indica. Journal of Apicultural Research*, **19**, 6–16.

Yadava, R.P.S. and Smith, M.V. (1971) Aggressive behaviour of *Apis mellifera* L. workers towards introduced queens. I. Behavioural mechanisms involved in the release of worker aggression. *Behaviour*, **39**, 212–26.

Young, L.C. and Burgett, M. (1982) Effects of synthetic 9-oxodec-trans-2-enoic acid on the foraging activities of honey bees. *American Bee Journal*, **122**, 773–75.

Zmarlicki, C. and Morse, R.A. (1963) Queen mating-drones apparently congregate in certain areas to which queens fly to mate. *American Bee Journal*, **103**, 414–15.

Zmarlicki, C. and Morse, R.A. (1964) The effect of mandibular gland extirpation on the longevity and attractiveness to workers of queen honeybees, *Apis mellifica*. *Annals of the Entomological Society of America*, **57**, 73–4.

Zoubareff, A. (1883) Concerning an organ of the bee not yet described. *British Bee Journal* **11**, 296–97.

AUTHOR INDEX

SUBJECT INDEX